Leaner Manufacturing

Lean Manufacturing, the production method first popularized by Toyota, has often been reduced to charts and tools by many US manufacturers – they've abandoned the original intent of the methodology. Many of the concepts and theories have become misconstrued, but systems such as Kanban (a Japanese concept used in Just-in-Time manufacturing) and applying core Lean principles to manufacturing remain valid today.

This book promotes *Leaner* Manufacturing™ as next-generation thinking involving the manufacturing process. It considers transformative tools that still work well, such as value stream mapping, and considers the manufacturing process as well as design theory.

Leaner Manufacturing™ simplifies current manufacturing terms including the four categories of manufacturing: casting/molding, machining, joining, and shearing/forming, as in layman's terms. This simplification creates an opportunity for invention, and the process for applying and securing a patent would also be included.

In addition, *Leaner* Manufacturing™ covers how to improve/modernize Kaizen Events (Rapid Improvement Events), Value Stream Mapping Processes, and Kanban practices as well as shows the power of using Agile Scrum practices and easier Lean Analytics.

Essentially, *Leaner* Manufacturing™ simplifies the valid concepts of Lean Manufacturing as well as manufacturing, in general, to make various processes easier and less expensive to implement. It shows how to make Kaizen Events (Rapid Improvement Events) more effective. And, it provides a way for leaders to ensure buy-in from their teams and stakeholders and provide for a motivated and inspired workforce.

Leaner Manufacturing

How to Make the Lean Production Process Easier, Faster, and More Cost-Effective

Terra Vanzant-Stern, PhD

Routledge
Taylor & Francis Group
A PRODUCTIVITY PRESS BOOK

Designed cover image: T&F

First published 2025
by Routledge
605 Third Avenue, New York, NY 10158

and by Routledge
4 Park Square, Milton Park, Abingdon, Oxon, OX14 4RN

Routledge is an imprint of the Taylor & Francis Group, an informa business

ISBN: 9780367775391 (hbk)
ISBN: 9780367775384 (pbk)
ISBN: 9780367775414 (ebk)

DOI: 10.4324/9780367775414

Typeset in Garamond
by codeMantra

This book is dedicated to the National Association of Manufacturers and the ASQ Manufacturing Division.

Contents

Preface

Lean Manufacturing can help your business post-pandemic, but *it can also help* you excel in an environment challenged by supply chain, leadership, and staffing issues.

Lean manufacturing has solid benefits, from increased profits to improved efficiency. But as we move to a post-pandemic world, it can be difficult to implement. This is why it is important to keep all the great concepts and tools promoted by traditional Lean Manufacturing but, at the same time, find a way to make it easier, essential *Leaner*.

This guide will explain how we have traditionally and successfully applied Lean Manufacturing principles and include ways to make the training, implementation, and maintenance smoother.

What Is Lean Manufacturing?

Lean manufacturing is a production strategy that reduces waste of time and materials. It uses techniques like Just in Time (JIT) and Kanban to make production more efficient. Kanban is a tool to visually track how much inventory is needed and where it needs to be. Just in Time means you only produce what you need when needed, reducing wasted time and resources.

The benefits of Lean Manufacturing are well established. Lean Manufacturing is a system that helps businesses become more efficient and profitable by identifying and removing waste from the process. It has been around for over half a century. Several studies have shown that companies that implement Lean Manufacturing approaches can bounce back within 6 months. The most famous incident in recent history is the Japanese recovery from the 2011 Tsunami.

What Is Leaner *Manufacturing™?*

Leaner Manufacturing™ (*Leaner*) recognizes all the grand schemes and tools offered by Lean Manufacturing. It is also a methodology that acknowledges and embraces the basic principles of JIT and Kanban. A focus on reducing waste and articulating the return on investment (ROI) of waste-saving activities is also a concentration. The difference between Lean Manufacturing and *Leaner is that it* is about increasing employee comprehension and, therefore, increasing the successful application of the methodology.

Lean Manufacturing is an organizational tactic emphasizing the importance of reducing waste and increasing customer satisfaction. Post-pandemic, it can help businesses by focusing on the following three elements:

1. **Identify the Bottlenecks in the Processes**: Bottlenecks cause inadequacies, which result in delays or increases in production costs.
2. **Eliminate All Non-Value-Adding Activities**: Non-value-added activities (NVAA) involve work that consumes resources but does not add value to the product or service.
3. **Focus on Quality First**: Quality checks should be visible and built into every production stage, not inspected at the end. Quality control checks should happen as products move through each stage of production.

The Future of Lean Manufacturing

Research implies that the pandemic's ramifications can continue for the next decade. In the post-pandemic era, there is a strong possibility that there may be less demand for new or expensive items and more demand for practical products such as clothing or food.

The issue is not whether Lean Manufacturing principles should be embraced during this extensive recovery period but how Lean Manufacturing tools and concepts can be made easier to understand and faster to implement.

While there are other manufacturing methodologies, Lean Manufacturing is the most progressive and articulated approach. However, employers looking for an easy route often do not adopt the practice.

Leaner Manufacturing™ offers theoretical minds an easier way to express the methods in smaller doses, allowing the employee to understand and utilize them.

The true power of *Leaner* Manufacturing™ includes understanding the basics of process improvement along with complementary methodologies such as Agile, Six Sigma, and basic Manufacturing Project Management. These topics will be briefly discussed to show the synergies and updated thinking and tool usage when it comes to *Leaner* Manufacturing™.

Terra Vanzant-Stern, PhD

About the Author

Terra Vanzant-Stern, PhD, PMP, SPHR/GPHR, is a Six Sigma Master Black Belt with experience in Manufacturing, Supply Chain, and Leadership. She is a popular host on IT PRO TV (ACI Learning) and a principal of SSD Global Solutions – **S**mart, **S**imple **D**ecision-Making. She is the creator of *Leaner* Six Sigma® and *Leaner* Manufacturing™ in addition to the Lean and Agile Project Management® curriculum.

About SSD Global Solutions

SSD GLOBAL SOLUTIONS, INC., Smart, Simple Decision-Making (SSD), began in 1996 as an international global mergers and acquisitions firm. In 2004, many companies that had previously invested in growing their businesses decided to tighten their belts due to the economic climate. This was an excellent opportunity for SSD to introduce Lean Six Sigma to their existing customers!

SSD remained committed to making Lean Six Sigma easier for students to understand and achieve success. The company's principal, Terra Vanzant-Stern, PhD, served as the ASQ Lean Enterprise Division Chair and met many interested in simplifying Lean Six Sigma concepts. Dr. Vanzant-Stern embarked on a mission to make Lean Six Sigma easier to understand and use. As a result, Dr. Vanzant-Stern has written several books and white papers on critical thinking, problem-solving, manufacturing project management, and leadership, available on Amazon.com and at major bookstores.

SSD's primary products are now all based on *Leaner* Six Sigma (L*r*SS©).

Visit www.SSDGlobal.net to learn more about Dr. Vanzant-Stern and her Lean journey, as well as our award-winning training programs.

Chapter 1

Understanding Process Improvement?

A quick understanding of Process Improvement and Lean Manufacturing is necessary to understand *Leaner* Manufacturing™ (*Leaner*) and its value to manufacturing environments. This chapter is primarily dedicated to process improvement. The next chapter will explain Lean Manufacturing as we know it now.

Process Improvement aims to identify an existing process that can be made better, faster, and more cost-effective. The term "better" is an umbrella term for quality and cost-effectiveness that considers all the financial aspects of developing the product.

In manufacturing, the emphasis is almost always on speed. This is true even without applying traditional Lean Manufacturing principles. In this context, faster means more efficient. Speeding up a process without considering the efficiency factor can be counterproductive.

Manufacturing and process improvement work in tandem. This includes adding processes and procedures to promote compliance, efficiency, and quality.

Process improvement is the business function of streamlining operations and ensuring the company remains competitive with others in the industry.

Process improvement is the ability to change the way a business operates. It is a systematic, continual process to increase the quality, cost, and delivery of products or services. The benefits of process improvement are endless.

Streamlining a manufacturing process can lead to many benefits. The first and most obvious benefit is saving money and time. When costs need to be

DOI: 10.4324/9780367775414-1

kept down, each minute or dollar counts. A streamlined process also means more control over the product since the work is done more efficiently. Implementing systematic and continuous improvements can measure the impact of quality, cost, and delivery improvements.

A process improvement strategy ensures a company remains competitive in the market, reducing business operations and streamlining operations.

The benefits of process improvement are plentiful and can include:

1. Improved quality
2. Increased productivity
3. Reduced waste
4. More customer satisfaction

When done correctly, process improvement is a powerful tool for addressing the challenges that manufacturing environments face.

The company makes changes in an unsystematic way; they risk making their operations even more inefficient than before.

Having a process that can improve efficiency and reduce waste is critical to the success of any manufacturing company. Process improvement, or kaizen in Japanese, has been used for a long time to eliminate waste and improve efficiency. Among the many ways to improve processes, one of the most popular is through Six Sigma management. Six Sigma management uses various tools to identify and improve manufacturing processes.

Process improvement, or kaizen in Japanese, is an essential component in any manufacturing environment. It has been used for many years to improve efficiency and reduce waste. There are many ways to implement process improvement, but one of the most common methods is Six Sigma management. This method utilizes a variety of tools designed to help companies identify and improve their manufacturing processes.

Indeed, quality and cost-effectiveness are considerations, but Lean Manufacturing is about being faster. This is one of the strong points of *Leaner* Manufacturing™. It is a return to evaluating quality and cost-effectiveness. *Leaner* also considers the value of Agile techniques, Six Sigma methodology, and traditional manufacturing project management.

Technically, in our evolving world, everything could be improved (higher quality). So, the *Leaner* approach would promote identifying an existing process or activity that, within reason, could be improved, faster, or more cost-effective.

The history of process improvement is essential. It provides necessary background information. The *Leaner* approach also makes the vast amount of information more accessible and understandable.

Lean Manufacturing is often referred to as a process improvement program. Other Lean programs, such as Lean Health, Lean Engineering, or Lean Manufacturing project Management, have already evolved into management systems and leadership development. This metamorphosis into management and leadership is another goal of *Leaner* Manufacturing™.

Process improvement (PI) is a theory about how to improve existing activities, make them faster, or make them more cost-effective. In other words, how can we improve the things we already do and hopefully keep improving them? There are several PI theories. They all have their own methodologies and tools and, in many cases, specific templates for purchase.

Some organizations and associations specialize in sponsoring a specific PI program. These groups have long histories, politics, and bodies of knowledge supporting that their way is the best, and these entities remain ongoing.

Eventually, however, most business professionals and leaders agreed that there was no one-size-fits-all plan. Not surprisingly, having a plan is always better than not having a plan. So, any PI program can have impressive results. Six Sigma, developed by Motorola, and Lean credit people, including Henry Ford, are among the top choices.

Today, a person may become certified in many PI disciplines through private organizations or universities. However, the trend is overwhelming to participate in training programs that combine Lean and Six Sigma methodologies. This combined science is known as Lean Six Sigma (LSS). Including the Six Sigma structure can only benefit Lean Manufacturing.

Lean Six Sigma, sometimes called Lean and Six Sigma, has tremendously increased Return on Investment. But almost two decades later, it is time for the next generation. We have learned ways to improve the core methodology and tools faster and cost-effectively. So, we are past due to an upgrade. *Leaner* Manufacturing™ is the next generation.

Using *Leaner*, students can apply models and theories on the fly without constantly consulting huge or complicated manuals. This is because the information is simplified in a way that makes sense the first time it is heard. Then, immediately, the notion is tied to an activity that is meaningful and easy to recall.

This abridged and accelerated approach allows the reader or student more time to work with the context of the issue instead of memorizing convoluted and sometimes outdated tools. With the new push in our society that less education may represent more time to reason things out by discussing issues, *Leaner* is a timely addition to the problem-solving and innovation arena.

Both Six Sigma and Lean Six Sigma are problem-solving models that allow maximum return on investment by improving, speeding up, or making existing activities more cost-effective using methodologies and tools designed for those specific purposes.

It is important to note that many organizations informally embrace the basic concepts. Many Lean and Six Sigma tools are now integrated into day-to-day business activities. Used together, Lean and Six Sigma increase the likelihood of success. Since Lean works to eliminate or reduce waste and Six Sigma works to eliminate or minimize mistakes, the methodologies are not mutually exclusive and work well together. This combined use in the industry is called Lean Six Sigma (LSS).

One premise of this book is that Lean Manufacturing benefits from using complementary methodologies and tools. Adding other thoughts and concepts makes Lean Manufacturing even *Leaner*.

Top Ten Tips for Manufacturing Process Improvement

Process improvement remains an important part of business strategy for companies in all industries. Whether it is a team or individual effort, there are many opportunities to improve processes that can save time and money while improving quality and safety outcomes.

Here is a summary of the top ten process improvement tips for manufacturing:

1. **Document of the Current Process**: Every company has a process, and it needs to be documented. The documentation can be used by new employees and those in charge of improving the process.
2. **Conduct a Process Flow Analysis**: Process flow analysis is a safe way to identify bottlenecks and inefficiencies in the production process and identify areas for improvement. The first step is to outline the process from start to finish and list all the steps involved. Next, assign a letter or number to each step according to its location in the process.
3. **Identify Opportunities for Improvement**: When improving a company's efficiency, the first place to look is in the manufacturing process. One tip is on how to identify and address opportunities for improvement:

1. Identify bottlenecks – Bottlenecks are components of the process that take up more time or resources than others. If these parts of the process can be improved, it will dramatically impact the efficiency of the entire process.

4. **Select the Right Tools and Techniques**: The right tools and techniques can help improve the manufacturing process. For example, statistical process control (SPC) provides a systematic approach to improving the methods by setting up controls and following the data. IT can also use the design of experiments (DOE) to optimize the processes. These two examples illustrate how using the right tools can help improve manufacturing processes.
5. **Involve Stakeholders in the Process**: Involve the stakeholders in the process. They will often better understand what needs improvement and can help generate a more comprehensive list. They must also be aware of how the changes will affect them and their employees. This way, they will feel invested in the process and motivated to see it through.
6. **Set Measurable Goals**: It is crucial to set achievable goals with the current resources. Use a SMART goal: specific, Measurable, Attainable, Realistic, and Time-bound.
7. **Pilot the Changes**: Pilot the changes on a small scale and see how they work. Always change them or abandon them if they do not work as planned. This is also a safe way to find out what needs to be changed, and it will provide more data for the pilot to work with.
8. **Implement the Changes**: This is the execution phase.

   Note: It can be difficult to get all employees on board with a new process. It is essential to keep the lines of communication open and stress the benefits they will see. Employees should feel like they have a voice and are listened to, but it is up to the owner or manager to make decisions.
9. **Evaluate the Results**: Traditional manufacturing metrics measure the effectiveness of process improvements. One option is to calculate the time required to produce one unit of product.
10. **Standardize the Improved Process**: Standardize the improved process by mapping out each step in complete detail, including what needs to be done in that step and when it needs to be completed. This chapter will first explain the basics of process improvement.

Modern technology is constantly changing the way we work and play. In today's world, people rely more often on technology than ever before. And in manufacturing, this has been no exception. With the help of innovative technologies such as computer-aided design, computer-aided manufacturing, and other digital tools, many processes have been streamlined to become more efficient and less time-consuming.

Regarding manufacturing, two types of improvements focus on sustaining and improving. Sustaining is keeping what is already in place, and improving is making changes to make it even better than before.

Lean Manufacturing Professionals who would like to understand better process improvement in general or the history of process improvement will benefit from studying material first presented within the Total Quality Management (TQM) framework. Works by W. Edwards Deming and Joseph Juran are still prevalent today. Other significant authors include Kaoru Ishikawa, A. V. Feigenbaum, and Philip B. Crosby.

The next chapter addresses Lean Thinking, a process improvement program like Lean Manufacturing that uses many of the same tools and methodologies. This chapter will also provide more information about process improvement.

Lean Thinking was originally a byproduct of Lean Manufacturing. However, the practice of Lean Thinking evolved, and many of the core ideas were not implemented back into Lean Manufacturing. *Leaner* Manufacturing™ capitalizes on the hybrid version of Lean Thinking and incorporates these practices into Lean Manufacturing.

Chapter 2

The Paradox of Lean Thinking in Manufacturing Practices

According to most accounts, Lean Manufacturing started before Lean Thinking became popular, as mentioned at the end of Chapter 1.

In 1991, the phrase "Lean Manufacturing" first appeared. The book "The Machine That Changed the World" by James P. Womack, Daniel T. Jones, and Daniel Roos of MIT mentioned the fundamental ideas.

Many people think Henry Ford was the first to adopt Lean practices. However, certain historians believe the name and idea originated in Venice in the 1450s.

The definition of Lean that is most widely used is:

"Lean is an idea or way of thinking that emphasizes cutting waste and simplifying procedures to save money, time, space, and materials."

The paradox is that Lean Thinking kept developing despite gaining popularity as a word and concept following Lean Manufacturing.

Popular discussions, keynote speakers, and publications have focused on the human component of Lean Thinking, how to use it, and how it differs from Agile and other approaches. As a result, several books are available demonstrating how to apply Lean Thinking to non-manufacturing sectors and startups.

Additionally, there are notable versions of Lean, such as Lean Office. Both Lean Manufacturing and Lean Thinking adhere to the following fundamentals of a Lean office:

- Devoted leadership
- Setting objectives and metrics

DOI: 10.4324/9780367775414-2

- Standardized procedures
- Work in progress (WIP)
- Good workflow
- An awareness of demand

In Lean offices, agile communication and a daily management system are used. They are usually visual, with labels on cabinets and signs indicating the path. Lastly, they promote a culture of ongoing development.

This chapter aims to demonstrate how incorporating Lean Thinking concepts, including undoubtedly hybrid variants of Lean Thinking – into standard Lean Manufacturing procedures might help Lean Manufacturing. This is where *Leaner* Manufacturing™ starts. Companies that already utilize Lean Manufacturing would not have to go through a steep learning curve to deploy Lean Manufacturing because they share some fundamental core beliefs and practices.

Lean Manufacturing, as commonly understood, started as a manufacturing technique focused on waste elimination. Its foundations were the Total Production System (TPS), first used by Toyota, and the Total Quality Management (TQM) tenets. To meet or go beyond customer expectations, TQM leveraged the participation of management, employees, suppliers, and customers. Lean Thinking streamlines the reading using a substantial portion of the original TQM content. For instance, this demonstrates the original five laws of Lean, which were first applied in Lean Thinking and then in Lean Manufacturing:

Zeroth Law: Since all other principles are based on this core idea, the initial law is zero. It reads, "The Law of the Market – Customer Critical to Quality defines quality and is the highest priority for improvement, followed by ROI (Return on Investment) and Net Present Value." Zeroth Law is a concept borrowed from science.

The first law is the Law of Flexibility. It asserts that "any process's velocity is proportional to its flexibility." In other words, the more adaptable and responsive the process is to changes, the better the manufacturing project implementation will proceed.

The second law, the Law of Focus, states that "80% of the delay in a process is caused by 20% of its activities."

The inertia of work in progress, or WIP, significantly impacts the velocity of manufacturing project implementation. The Third Law states, "The velocity of any process is inversely proportional to the amount of WIP."

The more unfinished tasks there are, the slower progress is due to various ground-level handicaps.

The bulky nature of products goes against the core principles of Lean, according to the Fourth Law: "The complexity of the service or product offering adds more non-value, costs, and WIP than either poor quality (low Sigma) or slow speed (un-Lean) process problems."

You can try to explain why passenger cars are becoming driver-friendly despite their complex engineering features and functions as an example of the 4th Law of Lean Six Sigma. Manufacturing principles: The bulk, complex manufacturing process, and product and service specifications render the offerings redundant.

The endeavor to simplify these five rules was one factor that helped Lean Thinking become more accessible and understandable. Before its current hybrid version, Lean Thinking attempted to simplify other notions.

The following things are regarded as the worst types of waste by both Lean Manufacturing and Lean Thinking (TIM WOODS):

- Transportation
- Inventory
- Motion
- Waiting
- Overproduction
- Overprocessing
- Defects
- Skills

Continuing with the idea that there are more similarities than differences between Lean Manufacturing and Lean Thinking, the following are the usual tools used:

- 5S
- Constraint Management
- Error proofing
- Kaizen
- Kanban
- Rate of inventory turnover
- Takt time
- Theories of just-in-time (JIT)
- Value Stream Mapping

As this country shifts its business style from control to management to leadership, the people doing the work can most often identify the changes needed to improve quality. Leadership must listen to and implement changes rather than dictate the solutions. The new *Leaner* Manufacturing™ tools emphasize continuous improvement as a guiding principle. The road to quality begins with minor incremental improvements. Significant sweeping changes rarely work. Some practical examples of Lean projects are:

1. Increasing the precision of forecasts
2. Lowering the number of orders that are denied
3. Increasing the cycle time for consumer loans
4. Cutting down on engine installation time
5. Reducing errors in an operating room
6. Lowering the rate of errors in pharmacy dispensing
7. Increasing employee hazard recognition's efficacy

The Deming wheel or cycle is a well-liked, simple-to-understand manufacturing project management technique. Lean Manufacturing and Lean Thinking support the plan, do, check, act (PDCA) paradigm.

- **Plan**: Find a chance and make a change plan.
- **Do**: Start small when implementing the change.
- **Check**: Examine the effects of the changes using data to see if they had any impact.
- **Act**: If the modification was effective, expand its implementation and consistently evaluate the outcomes; if not, repeat the cycle.

Both Lean Manufacturing and Lean Thinking use these visual aids and ideas:

- Control diagrams
- Pareto charts
- Mapping
- Analysis of the root cause
- Statistical Process Control (SPC)

Six Sigma is regarded as a distinct process by Lean Manufacturing and Lean Thinking, but the Define-Measure-Analyze-Improve-Control (DMAIC) model is valid.

The DMAIC model supports these constructs:

1. Establish process improvement objectives that align with business strategy and customer needs.
2. Assess the existing procedure and gather pertinent information for comparison in the future.
3. Examine connections.
4. Based on the analysis, enhance, or optimize the procedure.
5. Control to guarantee that any deviations are noted and fixed.

Mapping Value Streams (VSM)

The Value Stream Map is a valuable Lean Manufacturing and Lean Thinking tool. To correctly interpret the diagram, the project manager must be familiar with the representations, which are different from those used in flowcharting and include some intuitive symbols and others that rely on standard pictures used to depict specific items or situations related to JIT Manufacturing. Lean Value Streams analyze the current state and design a future state for events that take a product or service from its beginning to the customer.

VSM is also used in planning, supply chain management, service-related businesses, and the medical field. It is an essential part of Lean Thinking.

Business Process Mapping (BPM), which helps businesses become more efficient, is a closely related idea. Using a clear and complete BPM map or diagram, an external firm can determine whether improvements can be made to the current process.

Business process analysis is part of the BPM because it evaluates and contrasts a particular goal with the organization's objectives to ensure that all procedures follow the company's values and competencies.

Lean is a production practice with the crucial tenet of preserving value with less work. Operations that need to create value for the end customer are deemed wasteful. VSM aims to reduce production time and costs by eliminating waste and redundant processes. While most experts agree that VSM is a subset of BPM, some disagree and consider it a stand-alone activity.

Value Stream Mapping activities immediately benefit from productivity, mistake reduction, and customer lead times. By default, they also achieve the longer-term objectives BPM typically pursues, such as enhanced financial performance, customer happiness, and employee morale.

Both Lean Manufacturing and Lean Thinking accept the following definitions and two concepts:

1. Flow
2. Pull/Push

Flow

Flow is primarily concerned with reducing waste and enhancing flow by adhering to the Lean principles and a defined method for implementing each. If flow were a person rather than a concept, its objective would be to move products through a production system without separating things into lots.

Pull/Push

Supply Chain Management (SCM) aims to create a solution or supply for a goal, problem, or demand. The push and pull types of supply chain models are opposite in terms of the relationship between supply and demand; the push type is represented by make-to-stock (MTS), in which production is not based on actual demand, and the pull type is defined by make-to-order (MTO), in which production is based on actual demand.

Information technology makes it possible to change a production and sales business model from push to pull, which is one of the reasons SCM is so popular. Pull-type SCM is based on the demand side, such as JIT and a Continuous Replenishment Program (CRP) or actual demand assigned to later processes.

How Lean Thinking and Lean Manufacturing Differ

How Leaner *Manufacturing™ Can Help*

This chapter discusses the similarities and differences between Lean Manufacturing and Lean Thinking. A comparative analysis can help show how powerful *Leaner* Manufacturing™ is.

Remember that one of the critical distinctions between Lean Manufacturing and *Leaner* Manufacturing™ is that *Leaner* Manufacturing™ incorporates the aspects of Lean Thinking and Lean Manufacturing and embraces the evolution of Lean Thinking. This hybrid methodology has embraced tools rooted in Agile, Six Sigma, the Information Technology Infrastructure Library (ITIL), and the Capability Maturity Model (CMM), which Lean Manufacturing does not acknowledge or incorporate into the methodology. *Leaner* Manufacturing™ uses a hybrid approach to recognize the best process improvement tools regardless of whether they were initially part of the Lean Manufacturing toolbox. Additionally, *Leaner* Manufacturing™ simplifies the tools and concepts used in traditional Lean Manufacturing.

Lean Manufacturing is an industrial process that focuses on increasing a company's performance by using time, money, and materials to decrease waste and produce higher-quality products as efficiently as possible.

Although Lean Manufacturing and Lean Thinking share the same objective of increasing the company's efficiency, there are several critical distinctions between the two that anyone unfamiliar with either topic should be aware of.

Lean Thinking focuses on decreasing waste across the entire business, but Lean Manufacturing concentrates on removing waste in manufacturing. The most apparent distinction between the two is that Lean Thinking encompasses everything, whereas Lean Manufacturing solely involves production.

Lean Thinking considers long-term solutions for sustainable success, while Lean Manufacturing focuses on short-term improvements. Lastly, in terms of implementation, Lean Thinking is less concrete and needs to provide specific guidelines or techniques, just as lean manufacturing does.

While increasing efficiency may not be as crucial as finishing the work, Lean Manufacturing may be the most excellent option for doing so immediately without making long-term plans.

Lean Thinking is worth investigating if the objective is to achieve a broader range of improvement opportunities with a more holistic emphasis on sustainability and problem-solving. For instance, it might be simpler to implement short-term changes like rearranging machines or reevaluating inventory.

As mentioned, Lean Thinking could be more tangible and involve abstract concepts. However, this can also help because it allows you to address issues across the board rather than just one aspect of the business. Therefore, it might be worth investigating if it offers a broader range of improvement opportunities emphasizing sustainability and holistic problem-solving.

The first significant distinction between the two is how they address the problem: Lean Manufacturing emphasizes minimizing waste by doing more with less, while the other focuses on production rather than customer pleasure or service. Both approaches are based on continuous improvement.

However, Lean Thinking considers all aspects of a company's processes and determines whether they should be improved or eliminated. For instance, if an old factory needs to be updated to produce higher-quality goods, Lean Manufacturing will weigh the cost of the updates against the current output. If they are still determining if they are making any money off what they are producing, they will invest in improving efficiency to reduce costs.

Before deciding what to do next, Lean Thinkers consider the value they are bringing to their customers. Lean Thinking considers all departments within a company, including marketing, customer service, research and development, and others, rather than just production.

Lean Thinking, also known as Lean philosophy, is an approach to problem-solving that emphasizes innovation with existing products and processes rather than just trying to cut costs on everything. Lean Manufacturing, on the other hand, is a system of production that uses fewer resources, less space, and less time to produce a product. This method was first developed in the automotive industry in the early 1900s to create more efficient car production, and many industries have adopted it.

The critical distinction between these two strategies is that Lean Thinking concentrates on redesigning products or developing new ones to eliminate waste (i.e., overproduction). In contrast, Lean Manufacturing focuses on increasing efficiency by streamlining production methods while utilizing conventional tools and processes.

Lean Manufacturing's waste reduction and continuous improvement concepts have continuously produced value. However, using traditional Lean principles alone might not be enough in today's technologically complex and fast-paced industry. Incorporating contemporary tools and technologies can increase the advantages of Lean procedures, making operations more effective and flexible.

First, it is crucial to find the areas where modern tools can make the biggest difference. Data analytics, for example, can reveal inefficiencies in production processes that may not be apparent using traditional methods and offer actionable insights. Advanced analytics aid in making well-informed judgments, supporting Lean's emphasis on productivity and waste reduction.

Leaner Manufacturing™ supports combining the evolved components of Lean Thinking along with modern tools that can be used to improve Lean Manufacturing. It also allows process improvement tools to be incorporated from other programs such as Agile, Lean Six Sigma, CMM, and ITIL.

Chapter 3

Project Management in Manufacturing

Chapter 2 discussed how Lean Thinking has evolved into a hybrid methodology. Lean Manufacturing has yet to recognize these changes formally. The same is true of essential project Management. Updated project Management techniques could significantly contribute to making Lean Manufacturing *Leader*. Revisiting project management for modern advancements ensures attention to best practices.

Most archaeologists agree that the Great Pyramid of Giza, completed in 2570 BC, is the first evidence of project management. This project included the degree of planning, execution, and control involved in managing it.

Fast-forward to 208 BC, when the Great Wall of China was constructed. Although considered a world wonder, more documentation about this project is available. The labor force consisted of soldiers, ordinary people, and criminals. In the end, Emperor Qin Shuang ordered millions of people to finish this project.

It was not until 1917 that the idea of the use of a standard set of tools might be helpful in project management. It started with the Gantt chart, developed by Henry Gantt. This demonstrated a significant innovation in the 1920s. With anticipation, the Hoover Dam project began in 1931. Gantt charts are still in use today.

After the Hoover Dam project, project management as science developed rapidly via these critical events:

DOI: 10.4324/9780367775414-3

- 1956: The American Association of Cost Engineers (now AACE International)
- 1957: The Critical Path Method Invented by the Dupont Corporation
- 1958: The Program Evaluation Review Technique (PERT) Invented for the U.S. Navy's Polaris Project
- 1962: United States Department of Defense Mandate the Work Breakdown Structure (WBS) Approach

Progressive concepts and ideas developed along with project management in manufacturing and other fields. Dr. Eliyahu Goldratt published a book titled *The Goal* in 1984. Dr. Goldratt discussed the Theory of Constraints (TOC) in this book. He assumes that there is always at least one restriction. The TOC process determines the limitation and looks at ways to get around it. Critical chain project management is based on TOC techniques and algorithms.

Scrum gained recognition as a project management methodology in 1986, and PMI published the first edition of the PM BOK® in 1987. Three years later, the PRINCE technique became a reality in 1989.

Project managers are surprised to learn that there are three main accepted bodies of knowledge for project management: PM BOK®, PRINCE2®, and ISO 21500.

The Project Management Body of Knowledge (PM BOK®) is a US-based program supported by the Project Management Institute (PMI). It provides a set of standard terminology and guidelines. Although it overlaps with practices used in general management, unique thoughts such as critical path and work breakdown structure are not typically discussed in other management disciplines, such as financial forecasting or organizational development, offered individual certification programs.

Projects in Controlled Environments, version 2 (PRINCE2®), is a program that began as a joint venture between the UK government and Capita, a private company. PRINCE2® focuses on dividing projects into manageable and controllable phases. It encompasses Quality Management and offers individual certification programs.

ISO 21500:2012, Guidance on Project Management, is an international standard developed by the International Organization for Standardization (ISO). An interesting piece of trivia is that work began on this standard in 2007. This standard is the first of a planned series and aligns by design with other more established standards in areas such as Quality Management Systems (QMS) and Risk Management.

The Project Life Cycle (PLC) concept is common to all three primary bodies. It refers to activities necessary to fulfill project goals or objectives and is commonly known as the Project Management Life Cycle (PMLC).

The Project Management Life Cycle has six phases: Initiation, Planning, Execution, Monitoring, Controlling, and Closure.

The term "initiation" refers to choosing the objective to summarize the steps or phases in the cycle. Estimating time and resources, deciding on the job sequence, establishing the project's execution schedule, and conducting a risk assessment are all part of the planning process. Simply put, execution is the act of carrying out the tasks. At every stage of the project, monitoring and control take place. It entails monitoring the project's overall status, hazards, quality, and resources. Closing is the phase that offers all the tasks required for the project office to conclude the project effort.

Project Management begins with closely observing each phase of the PMLC and envisioning opportunities. This is a summary of the phases and tools.

Project Initiation

The Project Initiation Phase is the most crucial part of the PMLC. This phase establishes the scope. A significant outcome is the project charter, which starts with creating a business case. The following steps involve a feasibility study and determining the team members.

A Supplier-Input-Process-Output-Customer (SIPOC) analysis could be helpful in either the business case or feasibility study. This would identify all the stakeholders in the project and consider the nonhuman resources that may contribute to its success. Using a template to create the project charter is another uncomplicated way to make the process Lean. The Plan-Do-Check-Act (PDCA) methodology may be the best way to establish the PMO.

Before a project is initiated, a project must be chosen. This is the beginning. However, the PM is not frequently involved in this process. The selected project manager is given the assignment.

The PM delegated a project during the selection process and would still benefit from the information presented.

One of the purposes of project management is to create a project purpose and a roadmap. Project Managers should know where they are going and how to get there. However, those making the project often have not taken the time to thoroughly consider the culture, environment, or resources necessary to complete the project successfully. Additional challenges exist when the project incorrectly aligns with the business goals and objectives.

Any project should be carefully vetted to determine whether it benefits the organization. How does this project fit with the company's overall mission? If a connection cannot be made, it will not matter if traditional project management uses Lean Manufacturing.

Fortunately, the project manager can adopt Lean Manufacturing components when selecting a project. In traditional project management, the PM is programmed to consider two types of thinking.

The first is the mathematical method (mathematical), referred to as constrained optimization. This method involves calculating several different mathematical factors.

The second method is the benefit measurement method (benefit). This method enables the PM to effectively compare the benefits and values of one project against another.

Both the mathematical and benefit approaches are practical; however, these methodologies are more robust when enhanced by certain aspects of Lean Manufacturing.

For example, in Lean, selecting a project by performing a pilot or running an idea through the plan-do-check-act (PDCA) model is not uncommon. The PM can perform a type of what-if scenario using the PDCA. This will allow smaller projects or ideas to be jump-started and others to be killed without much time investment.

Lean projects also consider factors not always apparent in the mathematical or benefit structure, such as the following:

- How will the project/process strategically impact the business objectives and outcomes?
- How can the project/process be simplified?
- How will the project/process align with customer issues and complaints?
- How will the project/process manage human and nonhuman resources?
- How will the project/process enhance the business's core competency?

The agile approach answers these questions by default. This is because employees speak to one another during Scrum and Sprint sessions and capitalize on collective intelligence. The Scrum and sprint sessions can inadvertently uncover worthwhile future projects that will be considered later.

In addition, agile works with a framework that is only sometimes strongly represented in traditional project management. Agile project selection considers factors that include the following:

- Organizational culture
- Management Philosophy
- Timelines
- Project breakdown
- Roles
- Ability to secure an effective cross-functional team
- Experience needed
- Delivery method

Project Planning

Significant aspects of successful project management come down to good planning. These diagrams discussed later in this work include:

- The KJ Method or Affinity Diagrams
- Interrelationship Digraph (ID)
- Tree Diagrams
- Prioritization Diagrams
- Matrix Diagrams
- Process Decision Diagrams
- Activity Network Diagrams

Project management tools are available in this phase. For example, suggesting a 5S model, a five-step method of organizing and maintaining a workplace, before beginning the project may help in the execution if the environment is physically disorganized. Gemba walks, a term used to describe personal observation of work, can promote a greater understanding of constraints within the work environment.

A key performance indicator (KPI) is a business metric that evaluates factors crucial to an organization's success. Although it can vary from organization to organization, a strongly promoted KPI can be a potent driver of behavior.

The planning phase is the second phase of the project management life cycle (PMLC). The project manager enters this phase with pieces of information. The most important is the project charter (PC). The PC should summarize other documents created and reviewed in the initiation phase. The PC provides the intent of the overall project but not necessarily the direction.

How a project plan (plan) is constructed depends significantly on the PM's psychology and beliefs about project management theory. Traditional project management (traditional) supports a top-down approach in which decisions are made at the leadership level, and not all knowledge is shared with team members.

It is a plan-the-work and then-work-the-plan model, which does not allow detours. Agile techniques (agile) and, in some cases, Lean, on the other hand, promote the following:

- Transparency
- Frequent inspection
- Adaptation

In business models favoring transparency, everyone knows how the project is progressing. Based on the findings, team members who touch the product or service evaluate the process frequently and collaborate on adjustments. In agile, adjustments are made quickly.

As with other areas of project management, Lean thinking (Lean) embraces both traditional and agile aspects and makes decisions based on the project and the culture. It has often been remarked that Lean is more agile than agile.

The purpose of a charter for the PM is to get approval, funding, and resources for the project the PM wants to accomplish. The charter is also the first step in building the plan. In traditional and Lean, a PC (charter) is a document that outlines the proposed activity in an organized and controlled style. This charter is time-linear enough that tasks can be outlined immediately in a linear fashion.

The vision, mission, and success criteria should be considered.

Vision: The vision defines the project's "why." It is the higher purpose or reason for the project's existence.

Mission: This is the "what" of the project and states what will be done to achieve its higher purpose.

Success Criteria: The success criteria are management tests describing effects outside the solution.

PM BOK® Knowledge Area	PRINCE2® Components
Integration	Combined Processes and Components, Change Control
Scope, Time, Cost	Plans, Business Case
Quality	Quality, Configuration Management
Risk	Risk
Communications	Controls
Human Resources	Organization (limited)
Procurement	Not Covered

Figure 3.1 Work Breakdown Structure (WBS).

The plan should be attached to the charter to allow the PM to move forward quickly without waiting for separate approvals. The plan should be recorded using a work breakdown structure (WBS).

Work Breakdown Structure

Most traditional project managers will agree that using a WBS to build a physical project plan is the best approach. A WBS is a key project deliverable that organizes the team's work into manageable sections.

Templates are available to create a WBS, but it is a simple outline. WBSs are also used in education and to express ideas. For example, a textbook may show a WBS to demonstrate what is covered in the material (see Figure 3.1).

A WBS is also a speedy way to develop a project plan draft. It is a to-do list that categorizes everything that needs to be accomplished under a particular bullet item. The format noted in Figure 3.1 makes it easy to group similar tasks.

Using WBS helps the PM cost out and estimate time for each task. This is helpful when creating a cost or time baseline. The WBS breaks down functions to allocate resources efficiently to the specific task.

The Project Plan

In traditional project management (traditional), a worthy plan effectively balances time, cost, scope, quality, and expectations. A plan should incorporate the project's overall expectations, definition, schedule, risks to the organization, and the blueprint (list of activities). The philosophy and spirit of the plan should be captured in the charter. The charter influences the plan, and the plan records the specific activities along with their associated schedule, costs, and resource needs.

The plan shows the activities and indicates how they will be controlled throughout the project and any dependencies. Dependencies are tasks that may need to be accomplished before a critical activity. The PM always wants to stay aware of these situations.

Designing a project plan requires listing all the steps necessary for success. Each step is then assigned a resource, a timeline for completion, and an essential cost. Once the project plan has been reviewed, a time and cost baseline is made. This baseline is used from the beginning to the end of a project to determine if the project is within the acceptable parameters.

Project plans also need to consider constraints (things that could prevent project completion) and assumptions (things that are assumed to be in place).

Lean supports traditional concepts of charter and plan development in Manufacturing PM. However, in Lean environments, social factors tend to be considered when building a plan.

Agile techniques (agile) take a different approach to the charter and plan to develop a Manufacturing PM. Still, they support the Lean mindset that more buy-in and motivated employees are achieved if the PM considers the company's culture and values.

Core values are described as the essential and enduring tenets of an organization. These are the guiding principles that impact how the organization thinks and acts. Company ideology provides the glue that holds an organization together through time.

eBay, the famous online auction house, has five fundamental values that it posts on its website:

- We believe people are good.
- We believe everyone has things to contribute to.
- We believe an honest, open environment can bring out the best in people.
- We recognize and respect everyone as a unique individual.
- We encourage managers to treat others the way that managers want to be treated.

Values should be universal in the company and govern relationships with employees, investors, customers, and dealers. However, determining values is not enough. The entity must also be willing to invest in the necessary resources and tools and reward behavior that supports these values.

Companies list their values as things that set them apart, making them unique and special. Companies print and post their values. Most business values focus on the customer or personal integrity.

In Lean Manufacturing environments, the PM often functions in roles, sometimes participating firsthand in a project rather than just as a manager. Therefore, the PM is responsible for leading the group in supporting the company values.

Even when there is no ethical breach, the PM must be careful not to allow perception.

That behavior is unethical or not aligned with the company's core values.

Because most companies make decisions according to core values, the PM must also be aware of these values because values help people embrace positive change.

As noted earlier in this chapter, agile handles the charter and plan process differently than traditional or Lean.

Traditional project management, as well as Lean PMs, strongly emphasizes the WBS when developing a plan. Agile favors more ongoing conversation and communication than a specific defined plan and supports this with more graphical representations.

In traditional Lean, the phases of a traditional waterfall development Manufacturing PM force the PM to the next phase only when the previous one is complete. The plan reflects this. This is because documenting and adhering to the process in compliance-related issues is often more revered than the actual outcome.

The project often starts with a phrase called the problem statement. In some cases, it is a problem that needs to be solved. However, the problem statement can also declare what needs to be created or improved. Therefore, a problem statement might only sometimes meet a problem's criteria. The solution is the resolution, modification, or creation of a new process. A good PC has a well-defined problem statement as well as a solution.

A thought to consider is that traditional favors more structured approaches, whereas agile promotes creative, flexible, people-rather-than-process approaches. Lean methods tend to fall where in the middle.

Models for Planning

The following models can be used in initiating or planning. They are instrumental when the PC does not provide enough direction to use a WBS to determine the needed tasks.

Each of these popular prototypes also notes the type of project management most often associated with that model: traditional, Lean, or agile. Models are similar and overlap:

- Structured problem-solving traditional
- Plan-do-check-act (PDCA) model: traditional, Lean
- DMAIC: Lean
- People-Process-product/service method (3Ps): Lean
- What-why-where-who-when-which (6Ws) approach: Lean
- Machines-methods-materials-measurements-mother nature-manpower (6Ms) method: Lean
- A3 format: Lean Manufacturing
- Five why approach: Lean Manufacturing
- Six Thinking Hats®

Structured Problem Solving When Planning

Structured problem solving is popular and embraced by business management theories and is not limited to traditional. Although tools may vary slightly, the steps include the following:

1. Decide on which problem to pursue and define the problem.
2. Determine the root cause.
3. Generate workable solutions and choose the most likely one.
4. Plan and execute the solution.
5. Verify effectiveness.
6. Communicate and congratulate.

Project Execution

Initiation and planning are necessary for the productive execution of any project. Essential project management and Lean align with project management deployment strategies. Using a project plan with a work breakdown structure (WBS) and establishing metrics are good examples. What Lean offers for project management is a solid mistake-proofing aspect. The strategy to ensure the project's success often involves placing controls and detection measures within the project plan.

Visual feedback systems (Andon) may encourage quicker execution. Creating continuous flow eliminates waste and speeds up project processes. Andon means "Sign" or "Signal." It is a visual aid that highlights action items. For example, a flashing light in a manufacturing plant may indicate a line closure.

In this phase, Hoshin Kanri, a policy deployment tool, may help ensure consistent progress toward strategic goals. Hoshin Kanri is a method for ensuring that a company's strategic goals drive progress and action at every level within that company.

Project managers (PMs) agree that a project plan should be piloted or evaluated before execution. Traditional project management (traditional) and Lean thinking (Lean) have formal attitudes toward pilots and testing.

Leaner techniques (*Leaner*) have a different spin on how pilots and testing should be conducted. Project management life cycle (MANUFACTURING PMLC) approaches include the following information in the planning phase: Before a project is rolled out to the enterprise, it should be evaluated or piloted to ensure success. This concept is strongly emphasized in Lean.

Piloting and testing are ongoing at *Leaner*. Much attention is paid to the voice of the customer (VOC). *Leaner* can adapt a project plan to meet or exceed a customer's expectations.

Whereas this chapter will focus primarily on traditional and Lean methods and how they can work together or enhance the piloting and testing agenda, *Leaner* testing is briefly covered next.

Performance testing in *Leaner* does not align with traditional or Lean but offers thoughtful processes that traditional or Lean manufacturing PMs may want to incorporate. One such measure is quadrant testing. The *Leaner* testing quadrants (the quadrants) are based on a matrix Brian Marwick developed in 2003 to describe the types of tests used in extreme programming (XP) projects.

The quadrants make collecting information easier and do not follow a waterfall path. Quadrants are divided into these four categories:

Q1: Technology-facing tests that guide Manufacturing PM.

Q2: Business-facing tests that guide Manufacturing PM.

Q3: Business-facing tests that critique (evaluate) the product.

Q4: Technology-facing tests that critique (evaluate) the product.

Now, the quadrants are primarily used for technical projects. The quadrants consider themes. Two other terms, user stories and epics, must be considered when understanding the concept of themes.

The user story describes the type of user, what they want, and why. It helps to simplify a requirement. An epic captures a large body of work. It is a significant user story that can be broken down into several smaller stories. Testing quadrants primarily address a theme of a group of user stories that share a common attribute. For convenience, they are grouped.

The testing quadrant in *Leaner* is considered a standard way to evaluate and track performance themes but is only embraced by some *Leaner* practitioners.

A project pilot or a test before rolling out the project is to identify and manage risk, validate the benefit, and secure buy-in. Whereas all project management theories support piloting or testing, as noted earlier, traditional and Lean take a formal approach. However, all project management work emphasizes evaluating the pilot or testing strategy more than the pilot or testing strategy. Traditional supports the following steps:

State the goal based on the business purpose that is driving the project. Develop an implementation plan and solution scenarios and build prototypes.

Develop an evaluation metric and decide what measurements are necessary for piloting or testing to be considered successful.

Determine the timeline and cost baseline for the activity if the test will take more than 30 hours.

Administrative tasks also need to be considered when conducting a test. These tasks are related to staffing, schedules, and facilities.

Depending on the test's size and scope, resources will participate in the piloting or testing. Human resource selection may rely on statistical sampling in large-scale piloting or testing. To use sampling, a clear definition of the testing purpose must be established and documented.

Sampling is selecting units (e.g., people and organizations) from a population of interest. Studying the sample responses provides feedback on a project or idea's potential success or failure.

There is a type of pilot that differs from typical pilots and is used by both traditional and Lean. This is known as proof of concept (POC).

A POC is designed to demonstrate the feasibility of a proposed idea or concept to solve a business need. It may also be used during the project selection process. The disadvantage of using a POC is that it is so specific to the product or service that subject matter experts must be available to conduct a valid study. Also, there is a great deal of variation when conducting a POC. Because a POC aims to prove that a project or idea will work, one-off strategies specific to that product or service may be employed.

However, the basic steps for a POC are amazingly similar to those of a regular pilot, but they tend to contain more detail. Typically, a POC scheme would include the following:

Step 1: Define and develop
- Identifying stakeholders and team members
- Defining goals, inputs, objectives, scope, and success criteria
- Establishing resource commitments and finalizing a POC schedule
- Collaborate with stakeholders to prioritize functionalities
- Determine deliverables

Step 2: Engineer
- Configuring and assessing the required infrastructure
- Determining solution steps

Step 3: Execute
- Creating the test design for use cases and defining positive and negative test scenarios
- Designing and executing test scripts

Step 4: Evaluate
- Reviewing and validating the POC results with all stakeholders
- Comparing outcomes to success criteria and aligning implementation if appropriate

Once the PM has developed a solution to the problem statement, it may be tempting to skip piloting or testing. In projects involving software development, Manufacturing PM testing is not an option. However, this activity is commonly avoided in projects without an information technology (IT) component.

There are reasons for dodging the endeavor, but most often, the reasons are time, money, and scope. This is referred to as the triple constraint. The triple constraints are real concerns, but bypassing piloting or testing will only enhance these constraints and may lead to additional obstacles. Industry definitions explain time, money, and scope as follows:

Time: This refers to the actual time required to produce a deliverable.

Cost: This estimates the amount of money required to complete the project. Cost encompasses various things, such as the following:

- Resources
- Labor rates

- Risk
- Materials
- Scope

These are the functional elements that, when completed, make up the project's final deliverables. Lean and *Leaner* both offer ideas on piloting or testing schemes. For example, simulations can be modified to conduct pilot testing.

In addition, other Lean thinking and *Leaner* techniques (Lean and *Leaner*) methods can be adapted to evaluate a project's potential success and identify risks.

Before a project plan is executed, every attempt should be made to mistake-proof it. Mistake proofing is a Lean concept. In Japanese, it is known as poka-yoke. Poka-yoke means the use of automatic devices or methods that either make it impossible for an error to occur or make the error immediately obvious. A common example of poka-yoke is modern electronics, with connectors that are color-coded to the plugs.

The following mistake-proofing/poka-yoke steps have been modified to accommodate a project plan.

Create a high-level flowchart of the significant steps in the proposed project. Review each step, thinking about where and when human errors will occur.

For each potential problem, work back through the process to find its source and consider what can be written into the project plan to minimize that risk.

If possible, eliminate the step that is the root cause of the error. Make the corrective action easier than the error.

A closely related concept often included in mistake-proofing strategies and popular in Lean is inspection or observation. Inspecting and/or observing things, such as materials needed, human resources involved, and facilities, can also help evaluate a project in realizing and identifying threats.

A formal source inspection check before your proposed project activity takes place to ensure that conditions can manage the task is a quick but powerful way to set controls, especially if the controls are automatic and able to prevent the process from proceeding until conditions are right.

A less formal way to evaluate a project's success that also helps with communicating the project is thinking in terms of voices. The primary voices are the following:

- Voice of the customer (VOC)
- Voice of the employee (VOE)

- Voice of the process (VOP)
- Voice of the business (VOB)

Most Project Managers are aware of the importance of VOC. The VOE can offer a great deal of insight by simply asking certain employees to review the project plan draft. Often, the employee may be able to identify a risk the PM has not considered. The VOE is also helpful in identifying physical bottlenecks or undocumented processes.

When using VOP, the PM must carefully determine if the current process can manage the task recorded on the project plan. Finally, VOB, which stands for voice of the business but means voice of the industry, must be considered from a compliance standpoint.

Traditional project management, as noted earlier, offers tips on evaluation metrics.

Evaluation Metrics for Piloting or Testing

The metrics used to measure a pilot or test can also be used to measure the project's overall success. Performance metrics are the most important ones when assessing a pilot. These include, but are not limited to, the following:

- Schedule and effort/cost variance
- Resource utilization
- Change requests to the scope of work
- Number of problems reported by attendees
- Schedule and effort/cost variance

Schedule and effort/cost variance measure the performance as well as progress of the project against signed baselines. A timeline and cost baseline should be established if a pilot takes more than 30 hours to complete, not including the evaluation. A schedule and effort/cost variance are often called earned value in traditional project management. This metric integrates project scope, cost, and schedule measures to help the PM assess and measure project performance and progress. This metric uses past performance (i.e., actuals) to forecast future performance accurately. There are three main factors to consider:

- Planned value (PV)
- Earned value (EV)
- Actual cost (AC)

PV is the amount planned to be spent on the work or budget to accomplish the work breakdown structure. EV is the value of work performed to date. AC is the actual cost to date. Using these three variables, schedule and cost variance can be analyzed.

Resource Utilization

The next metric, resource utilization, measures resource productivity. Strategies for this are available, and they are transaction-based or observation-based.

Change Requests to Scope of Work

Changing requests to the scope of work should be extremely limited in the actual project. Still, it may be prevalent in pilots because pilots are flexible by design, especially the handful of pilots conducted in *Leaner* environments. Incorporating this metric can help determine where problems may occur in the project and can also be helpful when designing frequently asked questions or training.

Project Monitoring and Controlling

The Monitoring and Controlling process oversees all the tasks and metrics necessary to ensure that the approved and authorized project is within scope, on time, and on budget so that the project proceeds with minimal risk. Both Lean Manufacturing methodologies promote more people interaction than typical project management. Project management models rely more heavily on Gantt and other charts to track progress. Concentrating a little more on people's aspects will increase team accountability.

Lean specifically addresses Poka-yoke and Heijunka. Poka-yoke is the Japanese term for mistake-proofing, which involves eliminating possibilities for errors. An example would be color-coding a wiring template to assist the worker. The input and output would be the same color.

Heijunka is the Japanese word for level scheduling. This strategy aims to minimize disruptions caused by sudden changes in demand levels by matching the product family schedules with product-by-product schedules. To achieve the goals of level scheduling, the sales and production departments must agree on a fixed output volume and duration level.

Project Closing

The closing phase in the project management lifecycle confirms the completion of project deliverables to the satisfaction of the project sponsor and communicates the final project disposition and status to all participants and stakeholders. The concept of Standardized Work is often helpful during this phase. This is where documented procedures capture best practices. If Standardized Work is in place, it accelerates the closing process.

The PM BOK® specifically lists ten knowledge areas in project management addressed in a different structure within PRINCE2® and ISO 21500:

1. Integration
2. Scope
3. Time Management
4. Cost Management
5. Quality
6. Human Resources
7. Communications
8. Risk
9. Procurement
10. Stakeholders

There continues to be a bevy of new thoughts about effective project management. For example, the PM BOK® is in its 7th edition. However, the most intriguing thing is to impact essential project management.

Often, the management of projects is informal, and there may not be a specific process in place. Ways to improve project management in the workplace, even when a formal methodology is not in place, include:

- Establishing firm goals and objectives
- Structured planning
- Focusing on communication
- Determining the proper project management tools
- Identifying small "wins"
- Documenting essential items

This chapter discusses the distinct types of project management used in a manufacturing environment. Most depend on traditional project management techniques. However, each traditional project management technique has significantly updated its Body of Knowledge.

Project Management methodologies have more commonalities than differences, but it is always good to simplify and standardize the tools, templates, and approaches. The basic Lean Manufacturing process does not consider these modifications.

The PM BOK®, often considered the mothership of project management, makes changes to simplify the material and make it more usable. Thirty-three chapters cover every project management concept imaginable.

The latest update includes three significant changes:

- More clarity on the definition of a project.
- A section on non-work activities to better understand what their employees are doing at and out of work.
- A section on risks for project teams and how to manage risks during a project.

Addressing these changes could make the project management of the project *Leaner*.

This latest update to the PM BOK is a big one. The overarching goals of this update are to simplify, clarify, and better support project managers with their day-to-day work. This includes leadership and stakeholder engagement updates, resources and risks, team performance, and project closure.

When the best project management practices are not in place, costly overruns and delays can interfere with the bottom line. Even the reputation as a manufacturer could hurt relationships with clients and suppliers.

As a project manager, remember that the safety and health of the crew and team members are more important than anything. To help make Lean Manufacturing *Leaner*, keep these tips in mind for project management success.

Define the Scope of the Project

Every business, regardless of industry, will need to manage projects. Projects are necessary because they bring innovative ideas to life and solve problems. A project management methodology is the backbone of any successful project – all support defining the scope of work.

It is the part of project planning that focuses on completing goals. It includes a project's features and functions. Not taking time to do this or doing it inefficiently can cause trouble with the estimation. There are factors to consider before finishing a project.

Creating a Project Timeline

Accurate estimates of the time and resources needed to complete a project are essential. This will help obtain feedback from key stakeholders for realistic timelines.

1. Know the industry and company's standards for how long projects take.
2. Determine whether any factors outside of the control may affect the length of the project, such as changes in technology or legislative policy.
3. Determine whether this project is a one-person or team project.
4. Decide on the tracking tool and methodology. Example: Gantt Chart.
5. Plan the project stages, including objectives and metrics at each stage.
6. Estimate the time and resources needed at each stage.
7. Add all these numbers to determine how much it will cost and how long it will take.
8. Create a chart showing the start date, end date, estimated time, costs, and people involved.
9. Share the timeline with stakeholders, who must approve it before work begins.
10. Get feedback from them about anything that seems unrealistic about this timeline. Assign Tasks to Manufacturing Team Members.

One of the most critical aspects of project management is assigning tasks to team members. Ensuring each person knows what they need to do and when they need to do it helps keep the team on track and working toward their collective goal.

Monitor Progress and Make Changes as Needed

Good project management is essential to any manufacturing environment. Without it, the production process can be unpredictable and inefficient. The first step in good project management is monitoring progress and making necessary changes. Staying on top of deadlines, budgeting, quality control, and customer feedback is also essential. Honest communication throughout the company is critical to success as well.

Process Improvement (PI) programs depend on project management skills to execute and sustain improvement activities. Project management involves planning, organizing, securing, and managing resources to achieve specific goals. Today, *Leaner* Manufacturers must work to make these activities *Leaner* and more agile.

Internationally sensitive project management relies heavily on policy and leadership models the organization or culture supports. The *Leaner* Manufacturing Practitioner must also be aware of compliance issues governing a specific country's information technology (IT) activities. Most enterprise-wide process improvements will include an IT component.

Lean has always considered basic project management essential to successful process improvement. There are synergies between project management theory and Lean. Both essential project management and Lean are interested in establishing a sound plan, communicating with stakeholders, and conducting regular reviews. Likewise, project management and Lean are concerned with managing the schedule and containing costs. PMs are encouraged to use project management templates, especially those that focus on resource management and project execution.

The Lean model strongly encourages defining the problem, creating a clear picture of the current condition, and analyzing that condition before building a project plan for process improvement.

In other process improvement programs, a project plan is designed much earlier. A premium is often placed on the project plan as proof that work will be accomplished. The *Leaner* Manufacturing Practitioner needs to be aware of this expectation when working in various countries and with organizations that have established a formal Project Management Office (PMO).

Even in LSS, project management activities have room to be *Leaner* and more agile. For example, daily decision-making may not be delegated efficiently, cross-training opportunities may be overlooked, and a strategy to educate the team on LSS tools designed to improve critical-thinking skills may not exist.

Lean thinking begins with identifying and reducing waste. Waste may be defined as anything that cannot be connected to the customer's needs. A Lean project manager should use the Work Breakdown Structure (WBS) method whenever possible. A WBS outlines all the tasks that need to be accomplished. Primary tasks are given whole numbers such as 1, 2, or 3.

Subtasks are given designations such as 1.1, 2.1, or 3.1. Tasks related to the subtasks are recorded as 1.1.1 or 1.1.2. The theory is to break the task down to the most minor component to judge.

This method requires time and effort. It is also a quick way to identify any unnecessary steps in the process.

Lean project management requires being familiar with core documents related to project management best practices. These materials continue to evolve. For example, The International Organization for Standardization (ISO) has created a new international standard for Project Management, referred to as ISO 21500: Guide to Project Management. This standard is designed to increase global awareness of project management. The primary purpose of ISO 21500 is to enable multi-national organizations to coordinate their project management processes and systems. The *Leaner* Manufacturing Practitioner should benefit from this document since it strives to create a universal language around project management concepts.

Other respected documents that outline the project management approach include the Project Management Body of Knowledge (PM BOK®), supported by the Project Management Institute (PMI), and Projects IN Controlled Environments 2 (PRINCE2®) endorsed by the UK government are also internationally accepted collaterals.

The *Leaner* Manufacturing Practitioner should stay aware that, at this point, a process improvement effort becomes a series of small projects that must be executed. This requires using project management theory and tools. Familiarity with the new ISO standards for project management, as well as the PM BOK® and PRINCE2®, serves as a solid educational platform for *Leaner* Manufacturing Practitioners. The following section explains these documents, beginning with the PM BOK®.

PM BOK® management requires:

- recognizing that projects are temporary – there is a beginning and an end
- using a validated decision process to determine if a project should be done (risk assessment)
- consider factors related to timing, cost, quality, and resources
- developing and articulating project scope
- knowing how to close out a project and document any successes successfully

The PM BOK® has been recognized by the American National Standards Institute (ANSI) as an American National Standard (ANSI/MANUFACTURING PM I 99-001-2008) and by the Institute of Electrical and Electronics Engineers (IEEE) as the best guide for representing terminology and guidelines.

The PRINCE2® methodology encompasses project management, control, and organization. It involves a highly integrated approach that includes specific inputs and outputs. This process-based, structured methodology highlights how eight components, when understood and effectively addressed, can reduce risks in all types of projects.

In some cases, the *Leaner* Manufacturer professional will be working for a company that has already adopted a formal project management process. Lean strongly advocates working within any formalized program to improve it, speed it up, and make it more cost-effective.

The following information will provide an accelerated overview and highlights of ISO Standard 12500: Guide to Project Management, the PM BOK®, and PRINCE2®. There are more similarities than differences. This standard is scheduled for publication soon.

ISO 21500: Guide to Project Management states that for a project to be successful it should:

- Select appropriate processes to meet the project objectives.
- Use a Defined approach to develop or adapt the product specifications.
- Comply with requirements to satisfy the project sponsor, customer, and other stakeholders.
- Define and manage the project scope within the constraints while considering the project risks and resource needs to produce the project deliverables.
- Ensure proper support from each performing organization, including customer and project sponsor commitment.

It is vital to note that the content in ISO 21500, the PRINCE2® components and processes, and the knowledge areas in the PM BOK® are consistent. The PM BOK® is the most detailed account of project management terms and processes, PRINCE2® concentrates more on applied knowledge, and ISO 21500 provides a strong basic framework for a DMAIC project.

The PM BOK® supports nine knowledge areas:

All process improvement programs recognize that essential project management must be in place before process improvement may begin.

- Integration Management
- Scope Management
- Time Management

- Cost Management
- Quality Management
- Human Resource Management
- Communications Management
- Risk Management
- Project Procurement Management

PM BOK also states that the following phases are necessary for a successful project:

- Initiating
- Planning
- Executing
- Monitoring and Controlling
- Closing

PRINCE2® processes highlight the following activities:

- Planning (continued throughout)
- Starting up a project
- Directing the project
- Initiating the project
- Managing stage boundaries
- Controlling a stage
- Managing product delivery
- Closing a project

Other business activities in PRINCE2® include:

- defining the organizational structure for the project management team
- planning a Product/Service based approach.

Here is a Simple Comparison of the Main Topics in the PM BOK® Knowledge Area and PRINCE2® Components (Figure 3.1).

ISO 21500 is attempting to establish standards that can be adopted worldwide. PRINCE2® is a general framework, while MANUFACTURING PM I (Project Management Institute) focuses more on techniques. The big pluses of PRINCE2® are the following:

Other recognized forms of project management accepted methodologies include, but are not limited to:

- Traditional Project Management Method
- Process-Based Project Management Methodology
- Agile for Project Management
- Critical Chain Project Management (CMPP)

Traditional project Management is an approach that recognizes the same five developed Manufacturing PM components as the PM BOK®. It does not place emphasis on the knowledge area and is less flexible.

Process-based Project Management Methodology uses maturity models. An example is the Capability Maturity Model Integrated (CMMI®), a process improvement program developed by the Software Engineering Institute, part of Carnegie Mellon University. Another example is Software Process Improvement and Capability Estimation (SPICE, also known as ISO/IEC15504).

Agile for Project Management is seen as a series of small tasks. These tasks are flexible and executed on demand. Agile promotes being adaptive instead of being pre-planned. It also emphasizes human development and Manufacturing PM initiatives. Agile techniques are best used in small-scale projects. This creates a strong alliance with Lean.

Critical Chain Project Management is a method of planning and managing projects that emphasizes the resources required to execute project tasks. It was developed by Eliyahu M. Goldratt, who was well-known for his work on the Theory of Constraints (TOC) and his best-selling book, *The Goal*, from the 1990s.

Understanding the Project Management Life Cycle is valuable to the *Leaner* Manufacturer. Natural synergies exist.

Most project management methodologies/approaches agree that understanding the following areas is a key factor in successful project management:

- Integration Management
- Scope Management
- Time Management
- Cost Management
- Quality Management
- Human Resource Management
- Communications Management

- Risk Management
- Procurement Management

Projects defined as having a beginning and an end. There are activities a project manager uses to close out a project.

Although PMs use to manage their process improvement projects, there are companies that offer specific Lean project management software. For new project managers and process improvement practitioners, these packages can be useful for templates and navigation.

At times, the methodology must be simplified to make the project *Leaner*. The Plan, Do, Check, Act (PDCA) is a popular alternative to DMAIC. This methodology is also called the Deming Wheel.

- Plan: Identify Opportunity and Plan for Change
- Do: Implement the Change to A Small Scale
- Check: Use Data to Analyze the Results of The Change and Determine Whether It Made a Difference
- Act: If the Change Was Successful, Implement It on A Wider Scale and Continuously Assess the Results. If the Change Did Not Work, Begin the Cycle Again

Lean, Total Quality Management, and the PM BOK® endorse the PDCA model as a process improvement strategy. Although LSS prefers to use DMAIC, the *Leaner* Manufacturing Practitioner recognizes that it is not always reasonable due to time constraints or the simplicity of the problem. A PDCA may be used within the various phases of the DMAIC. A PDCA may also be used prior to considering DMAIC to determine the complexity of the problem. More information about the DMAIC model may be found in Appendix 3.

In other words, if quality and speed are critical to the customer, it will not be inexpensive. If the most crucial factor to the customer is low expense, and a fast turnaround, quality will suffer. This is a deviation from Lean thinking, although there are process improvement programs that may side with the PM BOK® on this issue. Therefore, the customer does not need to make this painful decision.

Another philosophy, where Lean and the PM BOK® differ, is the topic of scope creep. Scope represents the services that will be provided. Scope creep may be serious if a customer wants more service than was originally negotiated. Scope creep has the potential of distorting timelines and increasing cost. The PM BOK® is overly concerned about scope creep and strongly cautions against the practice. Lean recognizes that scope creep could be a problem but promotes always doing a little more than is expected. The purpose is to delight the customers with value-added activities.

When building a project plan, the *Leaner* Manufacturing Practitioner should stay cognizant and record any constraints. Constraints are conditions that may hinder project completion. For example, resource allocation might be a concern and therefore, be listed as a constraint.

Documenting constraints is the first step in building a risk-management strategy.

PMI Certifications

PMI offers several certifications. Each certification has different eligibility requirements:

- The CAPM® credential is designed for project team members and entry-level project managers, as well as qualified undergraduate and graduate students. The growing need for project management knowledge and proficiency bodes well for optimistic forecasts for IT practitioners.
- PM P® (Project Management Professional) – One of the most widely-recognized PM credentials requires the demonstration of a solid foundation of project management knowledge and practice.
- PMP™ (Program Management Practitioner) – This certification is geared toward those who manage multiple projects.

PRINCE2® Certifications

There are two PRINCE2® qualification levels: PRINCE2® Foundation and PRINCE2® Practitioner.

- PRINCE2® Foundation certification involves understanding the principles and terminology used in PRINCE2® and basic project management.
- PRINCE2® Practitioner is the highest level of PRINCE2® qualification. This certification verifies that a practitioner understands how to apply PRINCE2® principles when running or managing a project.

Cultural Awareness Training

Additional factors play a role in being a good project manager. These include:

- Change Management
- Leadership Development
- Measurement Systems Analysis
- Business Finance

In Lean and basic project Management, the term Change Management refers primarily to the processes and documentation in place to ensure the change takes place or that the process is implemented. Change Management may also refer to the training necessary to prepare employees for the specific change.

Leadership Development traditionally focuses on developing leadership ability. Within a Lean organization these methods are imperative to success but also include understanding diverse cultures, generations, and work ethics. Successful leadership develop Manufacturing PM is linked to:

- an individual's ability to learn
- quality and nature of the leadership development Manufacturing PM program
- genuine support for the leader's supervisor
- choosing the right projects
- choosing the right people
- following the right methodology
- clearly defining roles and responsibility

Measurement Systems Analysis (MSA) is a science that considers selecting the right measurement. It is necessary to study measurement as well as the measurement device. Are measures dependable and valid? What is the measurement uncertainty?

Statistics is the science of making effective use of numerical data relating to groups of individuals or experiments. Six Sigma and Lean have always included the field of statistics when measuring and analyzing data. Now, a stronger weight is placed on choosing the right software and making sure that the statistic is valid.

Business Finance plays a stronger role for the *Leaner* Practitioner. Buy-in and continued support of a project cannot be based solely on statistical data. Choosing the appropriate Return-on-investment formula and being able to measure project success, in financial terms, has become essential.

Moving forward as the *Leaner* Practitioner, it is imperative to remember that Lean is not just a matter of blending two phenomenally successful process improvement methodologies but rather encompassing a collection of bodies of knowledge.

In summary, stable project management plays a vital role in Lean. The *Leaner* Manufacturing Practitioner should stay aware that at point, a process improvement effort simply becomes a series of small projects that must be executed. This requires using project management theory as well as project.

Chapter 4

Agile Techniques to Make Lean Manufacturing *Leaner*

Leaner Manufacturing™ relies on an understanding of basic Agile constructs. An Agile Coach should be part of the team in a manufacturing environment that depends on Kanban. For those familiar with Agile, the primary difference in manufacturing project management is that items are released or delivered in Sprints in a non-manufacturing environment. Sprints are a prescribed amount of time the team works together.

In the Kanban environment, items are released as completed and evaluated. Many team experiences and meetings mirror Agile in IT and other non-manufacturing situations.

This chapter is designed to give those unfamiliar with basic Agile concepts an idea of how it works and how *Leaner* Manufacturing™ has borrowed from this methodology to speed up manufacturing processes.

The Agile Manifesto serves as the foundation for agile thought. A manifesto and a "Body of Knowledge" (BoK) are sometimes confused. A manifesto is more about purpose than details, even though there are similarities between what is offered in other fields, like project management, and what we consider a body of knowledge.

Seventeen independent-minded practitioners met in February 2001 and drafted the Agile Manifesto. Most of the participants had experience with programming. According to the Agile Alliance, many participants had varied ideas about Agile theory. They did, however, concur with four fundamental principles:

DOI: 10.4324/9780367775414-4

- **Individuals and interactions** over processes and tools
- **Working software** over comprehensive documentation
- **Customer collaboration** over contract negotiation
- **Responding to change** over following a plan

The Agile Twelve Principles were meant to be supplemented by these four ideals. It is still unclear how these 12 principles were first created. Nevertheless, these things give a clearer picture of the structure and goal of Agile. The following defines the universally accepted principles (The Agile Manifesto, 2001, Kent Beck, Martin Fowler, and Alistair Cockburn). *These are summarized to make them easier to understand*:

1. Our top concern is Ensuring customer satisfaction through timely and consistent delivery of essential software.
2. Accept requirements that change, even at the end of development. Agile methods are used.
3. Adjust to the client's competitive advantage. Provide functional software regularly, ideally within a shorter timeframe, ranging from a few weeks to a few months.
4. Developers and businesspeople must collaborate daily during the manufacturing project.
5. Assemble manufacturing initiatives around driven people. Please trust them to accomplish the job and provide the necessary environment and assistance.
6. In-person communication is the most effective and efficient way to share information with a development team.
7. The leading indicator of advancement is functional software. Agile processes facilitate sustainable growth.
8. The developers, sponsors, and users should be able to maintain a steady pace indefinitely.
9. Constant focus on useful design and technological excellence improves agility.
10. It is crucial to practice simplicity, which is the art of maximizing the effort that is not done.
11. Self-organizing teams produce the finest requirements, designs, and architecture.
12. The group considers ways to improve effectiveness regularly, then adjusts and adapts its behavior appropriately.

As discussed in the preceding chapter, Lean does not, in theory, have a manifesto or a set of principles accepted by industry. However, Edwards Deming's 14 Points from his book *Out of the Crisis* are the foundation for a substantial portion of Lean Thinking. Managers can apply Deming's points of management, which are 14 rules, to transform an organization's current management style into a more effective and optimized one. Each of the transformative concepts prioritizes quality over cost and stresses lead.

Recognized Certifications in Agile

Seeing how different people define the Agile Body of Knowledge (BoK) can be perplexing.

In general, a BoK serves as the foundation for any valid certification. However, Agile encompasses many approaches, including Just-in-Time Kanban principles, Scrum, XP, Lean, FDD, Crustal, and DSDM.

It is hard to prove a BoK that covers every idea and concept that drives Agile. This indicates that several unique certifications are available from private sellers.

Agile Alliance: The Agile Alliance was the first international agile community. Regardless of methodology, its goal is to support the advancement of agile principles and practices.

Scrum Alliance: A nonprofit professional membership organization, the Scrum Alliance works to increase awareness and application of Scrum. Several professional certificates are available from the Scrum Alliance:

- Certified Scrum Master (CSM)
- Certified Scrum Product Owner (CSPO)
- Certified Scrum Developer (CSD)
- Certified Scrum Professional (CSP)
- Certified Scrum Coach (CSC)
- Certified Scrum Trainer (CST)

Platinum Edge: Providers of training classes worldwide and developers of transition strategies and coaching for organizations moving to Agile Manufacturing project management.

AGILE Basic Tools and Techniques

Scrum is the most recognized word when Agile is applied to Manufacturing Project Management. It is an Agile methodology that can be used for any project.

Manufacturing project. Although initially used for software development, Scrum theory is now used in all business ventures.

Scrum

According to the Scrum methodology, every sprint should start with a quick planning session and end with a review. Scrum is a straightforward foundation for productive teamwork and a management and control method that cuts through complexity.

The manufacturing project manager, also known as the Scrum Master, will begin with a fundamental daily Scrum and ask the Scrum team three questions:

1. What did you do yesterday?
2. What will you do today?
3. Are there any impediments in your way?

Ideally, all individuals involved in the manufacturing project are part of the Scrum team. These gatherings occur in a circle, where everyone can stand facing one another. The position of team members encourages the notion that the session should be brief and accurate. Scrum sessions are conducted for individual manufacturing projects. Nonetheless, the Scrum meeting could support a daily report and reports on different manufacturing projects.

Concentrating on what each member did yesterday and what they will do today helps the team understand what has been done and what needs to be done. The Scrum master may occasionally consult a Burn Down Chart to facilitate this process. However, as we will see in the next section, this chart may also display Sprint's progress. For examples of Burn and Down Charts, refer to Figures 4.1 and 4.2.

The Scrum Master does everything he can to help the team perform at its best. Maintaining a healthy product backlog entails working with the product owner, setting up meetings, and removing any roadblocks to progress.

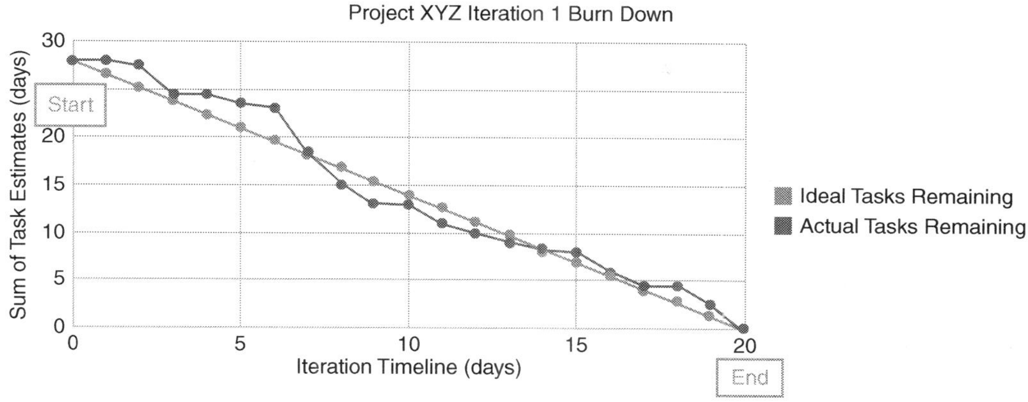

Figure 4.1 Burndown chart.

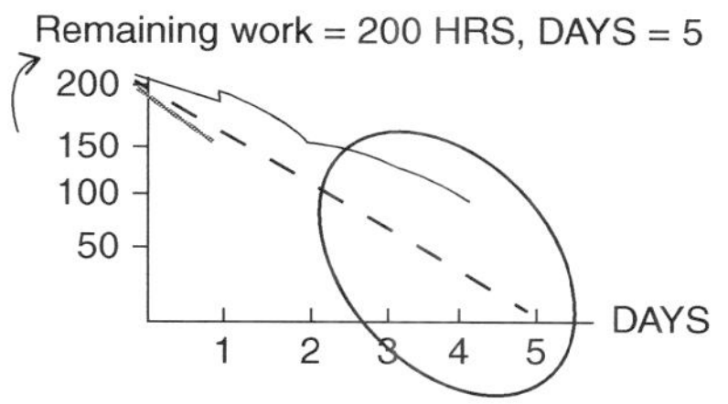

Figure 4.2 Burndown chart (manual).

Runs

In the Scrum version of Agile, sprints – also called iterations – are collections of work restricted to a regular, repeatable work cycle. These iterations can span anywhere from 1 week to 30 days, but they should all last roughly the same time. This increases the planning's accuracy and makes it simpler to remember the sprint timetable. The Scrum team works on highly focused and predefined tasks during this time. Nothing can be changed during the Sprint.

The first stage is a sprint planning session. Every individual who interacts with the process should be involved. Unlike daily Scrum meetings, which typically consist of individuals conducting the work who will report to other interested parties, if necessary, a Sprint requires the participation of all participants to be successful. Naturally, this is the time to decide on the sprint duration. Numerous factors, including the urgency of the manufacturing project and the availability of resources, influence the ideal sprint length.

Step two is to decide which item in the backlog needs to be worked on first. Because most sprints include more than can be completed, certain employees are less used to the process than others.

After reaching a consensus on the job, the objectives of the first Sprint are set, and each task is tackled logically and separately.

Quick Self-Assessments

The Scrum team convenes for a sprint retrospective at the end of each sprint to discuss what went well, what may be better, and how to apply any modifications. Typical discussion subjects are:

- What went well during the sprint cycle?
- What went wrong during the sprint cycle?
- What could we do differently to improve?

Agile Stages

Once you understand the terminology, it may be easier to think of the Agile process in terms of steps or stages.

Stage One: The Product or Process owner identifies the vision.
Stage Two: The Scrum Master or Product/Process Owner creates the roadmap, which is the high-level view.
Stage Three: The Scrum Master or the Product/Process Owner releases a plan identifying the timetable.
Stage Four: Product/Process Owner, the Scrum Master, and the Scrum Team plan the Sprints. Sprint planning takes place at the beginning of each Sprint.
Stage Five: During each Sprint, daily Scrum Meetings last no longer than fifteen minutes.
Stage Six: At the end of every Sprint, a report is given to the product/process Stakeholders.
Stage Seven: The Team holds a Sprint Retrospective.

Agile Manufacturing project Management promotes a value-driven approach that allows Manufacturing Project Managers to deliver high-priority, high-quality work. It is an iterative, incremental method of managing the design and building activities. It works well in engineering and information technology.

Typically, adaptive life cycles are iterative and incremental. The caveat is that the iterations in Agile processes are rapid – two to three weeks in length – with fixed resources.

Applying Agile techniques in the workforce takes various forms:

- Understanding the value of cross-functional teams.
- Scrum meetings designed to meet with the staff responsible for the manufacturing project daily.
- Looking at manufacturing project development as an incremental approach.
- Creating collaborative timeframes for manufacturing project items to be completed.

Agile in Manufacturing

Francois de Villiers watermarked Agile Manufacturing in his work, Lean Manufacturing World Class Manufacturing. Villiers said he never intended for his work to be released. He claimed to have written the handbook as a self-help manual for himself, but it did develop into this advanced explanation of lean manufacturing.

Villiers explains Agile Manufacturing as tools, techniques, and strategies that make it easier for plants or companies to manage unexpected changes. This could also include the capacity to reorganize operations quickly – and enter strategic alliances – to swiftly respond to emerging marketplace shifts and incorporate mass customization to satisfy customer needs. In a broad sense, adaptation is about the ability to react quickly to technical or environmental surprises. It is a way to thrive in a dynamic environment by managing complex inter- and intra-organizational relationships through new technological innovations, improved communication and business information strategies, organizational restructuring, and successful marketing initiatives.

Software development teams have used agile processes for years to develop better applications faster, but what about other industries? As it turns out, applying agile principles to manufacturing and supply chains can bring many benefits, so here is what you need to know about how to apply agile principles to manufacturing and supply chains.

Agile Manufacturing is a form of lean manufacturing that has emerged in response to manufacturers' changing needs and priorities. It is based on agile software development and relies on short, iterative design cycles, rapid prototyping, and continuous testing throughout the process.

The focus of Agile Manufacturing is communication and collaboration among team members. The team works together in iterative cycles, brainstorming ideas and creating mockups or prototypes.

Applying agile processes to manufacturing can help reduce costs, improve the quality of your product, and increase on-time delivery. Agile Manufacturing is an excellent way for companies of all sizes (even those with limited resources) to stay competitive. And for good reasons. With more than 50% of manufacturers using some agile method, the benefits are well worth the effort!

Agile Manufacturing aims to create a process that adapts easily and quickly to changes in customer demand. There are several ways you can implement this philosophy into your current production process, including:

1. Create a team with cross-functional members.
2. Implement sprints of work.
3. Continuously validate your product.
4. Reduce waste.
5. Allow for mistakes.
6. Encourage creativity.
7. Embrace transparency.
8. Become comfortable with change.
9. Get buy-in from management.
10. And have fun!

You need to know how the Agile Manufacturing process is going so that adjustments can be made as required. One way to measure success is by tracking the number of defects created daily. If there are too many defects, it could indicate that the team needs training on how to use the new Agile processes. Another way to measure success is by tracking overtime hours worked each week.

Chapter 5

Leaner Leadership in Manufacturing

While many of the secrets of *Leaner* leaders in manufacturing environments are widely known, these behaviors are not as well understood as they should be. This chapter will first explore the seven secrets of *Leaner* leaders in manufacturing environments through real-life examples and expert commentary to help develop a practical approach to improving business and working relationships.

- Put People First
- Respect Everyone
- Encourage Two-way Communication
- Promoting Continuous Improvement
- Focus on Results
- Lead by Example
- Be Persistent

Leaner leadership includes all the things mentioned in the basics of leadership. For example, like the general principles of leadership, the role of the *Leaner* leader in an organization is to provide direction and alignment with the organization. But the *Leaner* Leader's additional responsibilities also include:

- Aligning Purpose Process People and The Methodology to Build a Culture of Continuous Improvement

DOI: 10.4324/9780367775414-5

- Dealing With Issues Around Creating Change and Improving Business Results
- Introducing The Concept That a Process Improvement Is Not Worthwhile Without a Sustainability Component

Of course, *Leaner* leaders must be proficient in Lean problem-solving. This means thoroughly knowing the Plan-Do-Check-Act (PDCA) approach, discussed earlier in this book. *Learner* leaders should adopt a scalable and teachable approach to business operations.

- The *Leaner* Leader should understand and be able to relate basic lean practices. To summarize, they include:
- Identifying value.
- Mapping the value stream to identify and reduce or eliminate waste.
- Testing the flow to ensure the remaining steps run smoothly. A strong flow should have no interruptions, delays, or bottlenecks.
- Ensuring "Pull" is now being utilized instead of "Push." With improved flow, the time-to-market can be improved. This makes delivering products or services needed "just in time easier."
- Moving toward perfection.

The *Leaner* Leader needs to implement lean practices in a transformational way that supports organizational development alongside process improvement.

Typically, a Lean implementation involves an initial value stream mapping that defines the journey of the process that is a candidate for improvement. This can be done using Gemba, which follows the paperwork that leads to successful product or service delivery.

In this Gemba Walk exercise, the *Leaner* Leader needs to decide which people are the most appropriate to be invited on the walk. At a minimum, the team should include a subject matter expert and employees who work with the processes continually. Sometimes, it can also include the vendor supplier or senior management team member.

One of the biggest goals of *Leaner* Leadership is to ensure these goals are being met:

- Defining the service demand
- Understanding the lead time
- Matching up supply to demand
- Eliminating or reducing waste in the process, including lead time and cost

Defining the Service Demand

Identifying the wants of consumers and internal customers means paying attention to the voice of the customer (VOC) and the voice of the employee. During this process, it is essential to understand the specific business needs and expectations of the VOC and VOE.

Leadership in identifying service needs requires facilitation and motivation skills before considering implementation.

Does the service or product meet specific VOC requirements? Does VOE understand these requirements and have the processing power to meet them successfully?

The nature of the demand is a much easier concept to understand:

- What is the demand?
- How much time is needed to meet demand?
- How frequently does the demand need to be met?
- Who is requesting the demand and why?
- Where and when is the initial starting point?
- How will We Know the Demand has been Met?

The formula learns the actual demand and forecasts it is time to meet it. Businesses need flexibility to fit their schedules and ensure they have the resources to meet this demand.

How Frequently Does the Demand Need to Be Met?

Match demand frequency with the demands of an ever-changing market. This is a recurring need that businesses need to make a preliminary assessment of their ability to meet.

This demand is also reflected in other emerging customer expectations.

Who Is Requesting the Demand and Why?

A strong understanding of who is requesting the demand and why it is crucial to the success of *Leaner* leadership. It allows the leader to ensure they are working on the correct solutions and guiding and motivating the team in the right direction. By understanding the root cause of the demand, the leader can gear all efforts toward that result.

Where and When Is the Initial Starting Point?

Understanding location and timelines is crucial to meeting people in the right direction. It also eliminates any misunderstandings.

How Will We Know the Demand Has Been Met?

Understanding everyone's concept of success is essential before leading a manufacturing project.

Educating the workforce on how to engage in process improvement is extremely valuable. Lean Six Sigma recognizes the analytical business processes that need to happen but balances this approach with the recognition that everyone is the driver. Other process methods should be acceptable; however, Lean Six Sigma emphasizes the role of the leader as well as an understanding of change management and team dynamics. As used in business, leadership styles began to gain attention in the early 1900s. Over the decades, these styles have been repackaged and renamed. The original four most respected styles, because of research,

- Bureaucratic
- Charismatic
- Democratic
- Reactive

The most popular research shows that bureaucratic leaders are structured and follow established procedures. Charismatic leaders lead by inspiration. Democratic leaders are concerned with consensus. Reactive leaders make decisions immediately without considering feedback from other employees.

In 1939, a group of researchers led by psychologist Kurt Lewin identified only three leadership styles. This work was based on study groups of children and is still highly respected today. The children were assigned to one of the following groups:

- Authoritarian
- Participative
- Laissez-fair

The autocratic leader provides clear expectations, and there is an apparent division between leader and follower. The participative or democratic leader offers guidance to group members, but this type of leader also participates in the group, allowing input from other group members. The laissez-faire leader delegated most tasks to the team members.

Lewan's research indicated team members were the most productive in autocratic environments but more creative in participative environments. His research also stated that the least productive and creative environments were the groups designated as laissez-faire.

Lewan's research deemed the democratic leadership style as the most effective. He believed that although autocratic leadership appeared to offer the most productivity, it was the leadership style most likely to be abused. However, Lewin also believed that authoritarian leadership was effective when their deadlines did not allow for group decision-making and when the leader demonstrated that they were the most knowledgeable team members.

All research on leadership styles initially promoted the idea that everyone had a specific style. Much merit was given to the notion that a good leader must identify their core style to be enlightened and expand their leadership abilities. Newer thinking is based on situational leadership. This means that the leadership style would be dependent on the actual situation.

Leadership styles have been expanded to include the following:

- Relations-oriented leader
- Servant leader
- Transformational leader
- Task-oriented leader

Relation-oriented Leaders are also known as people-oriented leaders. In this style, the leader tries to support and mentor the team members rather than set the supreme direction.

Servant leaders have become a popular theory. This leader acts informally but takes a more informal approach and makes collective decisions by consulting the group. It is like democratic leadership. However, most publications indicate that this type of leadership is more common.

It is more sensitive than the democratic leader because this leader is concerned with how people feel and why they vote the way they do. Did, as opposed to just making consensus decisions.

Transformational Leaders motivate the team, giving them a reason to adopt a change. This type of leadership is like charismatic leadership, but the focus is not on the leader's traits or popularity.

Task-oriented Leaders are known for focusing on what the team needs to achieve. They are like result-oriented leaders but are more aware of the needs and welfare of the team members.

Most current leadership research suggests that all leaders must be more engaged to lead teams in the right direction. Leaders must act as catalysts for change.

Leaner Leaders seek "mission-critical manufacturing projects" and provide the management energy and horsepower to free up resources. But more than that, Lean Six Sigma Leaders have the vision to "imagine the future." An effective Lean Six Sigma leader identifies high-potential employees and understands the value of training.

Leadership is an essential quality that all manufacturing project managers are presumed to possess. However, the amount of actual ability is variable. Some acquire it naturally through inherited qualities, while others may benefit from formal leadership development.

A critical skill for leaders is the ability to manage their learning. The first step is to conduct self-assessment. Leadership development is a continuous process, not an annual event. It is crucial to adapt leadership styles to each person's needs and not be afraid to collect input from others.

The personal attributes and character of leaders are varied. The competencies (knowledge, skills, and abilities) a person needs to lead an organization at a given time depend on various factors. However, most people will agree that a good leader should possess common sense and judgment.

How an organization is structured often dictates the leadership role in a Lean Six Sigma manufacturing project. For example, organizations that favor matrix management can positively and negatively affect the Lean Six Sigma leader. In this type of company, a link between senior management functions and self-contained work cells is sometimes maintained through a matrix structure in which personnel assigned to functions are deployed temporarily to the cells. This approach maintains specialization's advantages while facilitating coordination within cells.

This arrangement is not easy to run. The managers of the various work cells may compete with their counterparts for the services, such as quality control and maintenance. Managers of functions may be concerned that temporary personnel assignments to cells may become permanent. No consensus on the ideal solution has emerged. However, some continuity of staffing may be needed for the success of the manufacturing project, and permanent assignments may impair the ability of the functions to maintain specialized knowledge.

As noted earlier, the new *Leaner* tools focus on continuous improvement as a guiding principle. The road to quality is paved with minor incremental improvements. Significant sweeping changes seldom work. As this country moves its business style from control to management to leadership, we find that the people doing the work can most identify the changes necessary to improve quality. Leadership must listen to and implement changes rather than direct them.

Leaner Leaders are encouraged to think in a new way and may be involved in any of the following activities:

- Acting as a catalyst
- Asking the right questions
- Creating a responsive manufacturing project solution
- Developing options and alternatives
- Discovering and exposing ideas
- Effecting timely decision-making
- Establishing effective and efficient manufacturing project start-up
- Establishing team ownership
- Expediting decision-making
- Facilitating the design team
- Identifying and clarifying organizational structure
- Improving communication
- Integrating customers into the total team
- Orienting the total team to a mutual goal
- Providing a framework by which to benchmark manufacturing project success
- Reaching design consensus
- Synthesizing ideas
- Testing options and alternatives
- Uncovering opportunities
- Understanding total requirements

In Western corporate organizations, leadership issues must align with vertical integration, horizontal diversification, growth by merger or acquisition, and shareholder interests. Lean Six Sigma encourages working in tandem rather than against the business system.

Leadership development differs in Lean because typical leadership programs focus on the leader's development, such as the personal attributes desired in a leader, desired ways of behaving, and ways of thinking or feeling. *Leaner* Leadership focuses more on the development of leadership as a process. This includes the social influence process, interpersonal relationships, and team dynamics. The advantage of this approach is that a person who may not feel like a natural leader may be more comfortable learning a process than changing their overall personality. Learn Leadership is also very flexible and supports changing strategy when necessary.

Leaders play a critical role during change implementation. Effective *Leaner* leadership will reduce the adverse reaction to change.

The *Leaner* Leader must understand the dynamics of companies with a Manufacturing Project Management Office since this entity will monitor all manufacturing projects.

Typically, the accepted functions of the manufacturing PMO are:

- To provide support and guidance to managers in manufacturing project implementation.
- Introduce the proper process and evolve a suitable methodology to develop acceptable standards.
- Design training programs to prepare team members to perform their tasks efficiently.
- Continuous monitoring of manufacturing project progress and mentoring the team members.
- Select, introduce, and oversee the use of appropriate software tools.
- Constitute a panel of program managers and train them to run multiple interrelated manufacturing projects.
- Timely resource allocation to complete work within specified time limits.

Choosing the right person to implement and lead the *Leaner* movement in the workplace involves selecting a leader who understands the importance of the scalability of the product or service and the need to motivate all parties involved. This person should have the ability not only to implement change but also to inspire change. Leadership that works is committed to continuous improvement. Commitment to continuous improvement requires managing change, which is the next chapter's topic.

Leaner Leaders might consider incorporating project management and Lean tools into leadership activities.

Making Communication Easier

A few things can be done to improve communication. Some of these items are better addressed in the planning phase, but often, people planning the project are not on the same team as those executing it, which is why they are covered in this section. The *Leaner* Leader can borrow from other methodologies focused on process improvement. There are also general communication tips that can be useful.

1. Let the project team determine, if possible, what tools they would like to use.
2. Communicate early and often.
3. Leave room for adjustments on your work breakdown structure (WBS).
4. Watch for opportunities and threats.
5. Keep a sense of humor.

During this phase, the specific activities include, but are not limited to, the following:

- Managing and tracking decisions
- Managing and tracking action items
- Execute and revise the project schedule
- Manage risk
- Managing and Tracking Decisions

The PM is responsible for ensuring decisions that need to be made are made before they impact the project, and decisions are placed in the repository of records for future reference.

In traditional project management, a decision-tracking log is often a common practice. This is a good start because a decision log is a spreadsheet that will provide the following:

- A reference for the decision
- Date decision made
- What was agreed and why?
- Who agreed to it?
- Where you can find more information or support documentation (optional)

In Lean decision-making, the emphasis is more on deciding on the document.

However, understanding how agile approaches a problem may make the decision log more robust.

With that in mind, here is the basic structure of an agile framework that should help you better handle tough business decisions:

- Start with eliminating irrelevant information or choices that do not align with your project.
- Anticipate decision anxiety by committing to a time frame for taking the first step. Limit how much time you can spend researching or contemplating a choice before committing.
- Set a date on the calendar to review critical decisions weekly.
- Be deliberate about when you involve outside feedback. Bring in collaborators and partners.
- Practice being present with the consequences of your good and bad decisions.
- Managing and Tracking Action Items.

The PM is responsible for ensuring that tasks too small to appear in the project schedule are recorded and completed.

In traditional project management, a list is kept, or items are added to the WBS. This can be enhanced by using a Gantt chart to see when tasks are being finished visually. Keeping your team on track is much more complex than recording the results.

Daily Scrum meetings, which are used in agile, hold meetings in which participants are asked daily to report the following along with their teams:

1. What did you do yesterday?
2. What will you do today?
3. Are there any impediments or obstacles in your way?

By focusing on what each person accomplished yesterday and will accomplish today, the team gains an excellent understanding of what work has been done and what work remains. The daily Scrum meeting is not a status update meeting in which a boss collects information.

About whom is behind schedule. Instead, it is a meeting where team members commit to each other.

These are a few examples of impediments or obstacles:

1. I cannot get the X group to give me any time, and I need to meet with them.
2. I cannot get the vendor's tech support group to call me back.
3. I need help debugging a problem with X.
4. I still have not got the software I ordered a month ago.
5. I am struggling to learn X and would like to pair with someone on it.
6. My X broke, and I need a new one today.
7. Our new contractor cannot start because no one is here to sign her contract.
8. The department VP has asked me to work on something else "for a day or two."

Execute and Revise the Project Schedule

Keep the project schedule updated by obtaining the status of project tasks and updating those tasks in the project schedule. The project schedule should be monitored and updated regularly.

This is the same for anyone working in project management. It is about frequent and valid communication. The advantage of lean and agile communication is that there is an underlying theme: input from the team is essential, and people are encouraged to ask questions.

Manage Risk

The project charter identifies an initial list of risks and management approaches. The project manager must monitor the risk list, identify any that have become issues, and implement the contingency plan.

Typically, a risk register and issue log are maintained in traditional project management. Like a decision or issues log, this helps manage the project and identify best practices.

Lean Six Sigma offers several tools to manage risk. It also expands on the concept of what is a risk. Risk is not always concentrating on what things could go wrong. Risk examines the internal environment, such as the following:

- Business vision
- Leadership
- Ethical issues
- Organizational commitment to human capital

When exploring risk, setting objectives and risk response need to be discussed with the team.

Additionally, the control and monitoring methods must be fully understood. Some tools include the following:

- Cause and effect matrix.
- Control charts to track the process.
- Failure mode effects analysis (FMEA).

Risk identification, monitoring, and resolution are critical to completing a project. Part of controlling a project during the execution phase is to have an established risk management process. This process is begun as part of project planning (see risk management subsection of the project planning phase section) and is kept current until project closes out. The critical elements of this process are the following:

Creating a central repository for risk information and associated documentation of risk items and resolution strategies

- Summarizing information on a risk form
- Including a risk summary in the regular status meetings

Providing a consistent and ongoing evaluation of risk items and development of risk strategies

- Identify the risk.
- Evaluate the risk.
- Define a resolution.

In all cases, risk management is an iterative process that is performed throughout the project.

The initial list of risks that begins with the project will evolve. Risk identification meetings should be held to ensure that new risks are added and resolved risks are eliminated.

Cause and Effect Matrix

The fishbone diagram is an easy but powerful cause and effect matrix. This tool is useful when determining the root cause or uncovering a bottleneck. It is also helpful when explaining why a process is not working.

Step 1: First, write down the exact problem you face. Where appropriate, identify the issue, who is involved, and when and where it occurs. Then, write the problem in a box on the left-hand side of a large sheet of paper and draw across the paper horizontally from the box. This arrangement, looking like the head and spine of a fish, gives you space to develop ideas.

Step 2: Determine the significant factors involved. Next, identify any other factors that may be part of the problem. These may include systems, equipment, materials, external forces, people involved with the problem, etc.

Step 3: Identify possible causes. Now, for each of the factors you considered in Step 2, brainstorm possible causes of the problem that may be related to the factor.

Step 4: Analyze your diagram. At this stage, you should have a chart showing all the possible causes of the problem that you can think of at that particular time.

Control Charts

A control chart is a graph used to study how a process changes over time. Data are plotted in time order. A control chart always has a central line for the average, an upper line for the upper control limit, and a lower line for the lower control limit.

One of the steps in FMEA is establishing a risk priority number (RPN). This is a numeric assessment of risk assigned to a process or steps in a process as part of FMEA. A team assigns each failure mode a numeric value that quantifies the likelihood of occurrence, probability of detection, and severity of impact. It is a number that typically looks at tables associated with severity, occurrence, and detection as a composite of the actual risk. Values in each category usually range from 1 to 10 but can range as high as 1 to 1,000.

Additional Tools

The Gantt chart is a widespread tool for monitoring WBS. It gives a graphic snapshot of your project's progress toward completing certain activities and tasks.

Lean thinking and agile techniques (Lean and agile) take a more severe approach to closing projects than traditional project management (conventional). This is due to the concept of continuous improvement. Lean and agile are particularly interested in documenting lessons learned and how the plan may have changed or been modified.

Project management has consistently promoted the definition of a project as something with a beginning and an end. Lean and agile believe in continuous improvement (CI). CI focuses on making the current project better, faster, and more cost-effective the next time a similar project is done or on making the existing project more vivacious.

CI models include plan, do, check, act (PDCA), but other models, such as Define, Measure, Analyze, Improve, and Control (DMAIC), are also in the mix. Kaizen, also known as CI, is a long-term approach to work that systematically seeks to achieve small, incremental changes in processes to improve efficiency and quality.

Because the concept of CI is that CI is the responsibility of every worker, not just a selected few, closing out a project involves more than the customary activities done when closing out a project, such as the following:

- Formal sign-off
- Releasing resources
- Updating paperwork
- Handling any procurement issues

These activities are important and well respected by Lean and Agile; however, many things considered friendly are not required in traditional project management. Lean and agile consider these activities essential:

- Celebrating success.
- Talk about lessons learned.
- Record best practices.

- Create standard work.
- Celebrating Success.

Working on projects is often hard work, requiring many hours of dedication. This is why it is important to recognize team contributions to make any project successful. Honoring success can happen at every level of the operation. The annual performance bonus, the uplifting speech at the holiday party, and the media release broadcasting news of a significant corporate achievement are all acceptable ways to recognize success. But it is also essential to entrench recognition of a job well done on a small scale. In general, celebrating project success motivates and produces happiness for the employees. Dopamine is released into our brain when we anticipate achieving something, or we do achieve it, and it feels good, so we want more. An excellent point to remember is that you must recognize your success if you want other people to do the same.

Talk about Lessons Learned

Talking about lessons learned is an excellent team-building activity and a practical project management tool. Organizations gain information through experience, which can be relevant to future projects or organizations and should be retained for future reference.

Identification, tracking, and consistent communication of lessons learned can save organizations time and money when embarking on similar efforts by preventing recurrence in other business areas or different projects.

Lessons learned are important because they may be used as tools to improve project management efforts in the future.

The items covered during this conversation should include, but not be limited to, the following:

- Identifying the event and subsequent lessons learned.
- Documenting the lesson learned and associated events.
- Acting to resolve or mitigate the situation.
- Communicating critical information to relevant stakeholders: when the event occurred, as it is mitigated, and post–project close-out.
- Maintenance or tracking of the lesson learned for future reference during other projects or by other personnel.

Record Best Practices

The best practice is a technique or methodology that, through experience and research, has proven to lead to a desired result reliably. It is similar but different from the standard work, which is discussed next. Best practice can usually be applied or modified to work with multiple projects or activities. Software is available to help collect and articulate best practices. Most Lean and agile practitioners use a spreadsheet or simple database.

Create Standard Work

Developing standard work is one of the more difficult Lean Six Sigma disciplines; however, if efficiently developed, it should allow anyone to perform the job without any variance in the desired output. It makes future work more efficient and includes a detailed definition of the most efficient method to produce a product (or perform a service) at a balanced flow to achieve the desired output rate. It can often be created at the end of a project.

Why Should the *Leaner* Leader Implement Standard Work?

Implement standard work to make it possible to identify and eliminate variations in the operator's work and to sustain gains achieved from improvement activities. It may also be used to provide a baseline for future improvement.

- Elements of Standard Work
- Total labor time
- Takt time
- Cycle time work
- Sequence
- Job steps
- Machine time
- Standard work in progress (WIP)
- Desired output
- How to Develop Standard Work

- Engage employees in the development process
- Encourage team members to collaborate
- Be realistic
- Guide through the creation and provide final approval
- Design with the intent to make problems visible
- Capture takt time for the process

Chapter 6

Change Management in Manufacturing

Over the past decade, the emphasis on Change Management has expanded. The roles of a fully developed management department might include:

- Advisor
- Analyst
- Consultant
- Coordinator
- Facilitator
- Lead
- Manager
- Director
- Vice President
- Practitioner
- IT Change Manager/Director

In some manufacturing and non-manufacturing environments, the role of the CMP is assigned formally or informally to the Human Resources Department or the Process Improvement group. The process improvement group can have several names, such as:

- ISO Department
- Process Improvement Department

DOI: 10.4324/9780367775414-6

- Quality Department
- Continuous Improvement Department

Many manufacturers, however, do not have a defined department to oversee change management, and therefore, often, the person in charge of ensuring that the manufacturing process is working successfully also assumes the role of change manager. Please note that when this chapter refers to Change Management Professional (CMP), although it is used as an umbrella term to make the Lean Manufacturing process *Leaner*, this chapter promotes that as a *Leaner* Manufacturing professional, it is necessary to assume some, if not all, of the responsibilities typically handled by the CMP.

Change related to *Leaner* manufacturing must be compelling. Integrating openness into the daily work life is essential in a *Leaner* organization. No one wants to be in a situation where change results from crisis. As demand changes, so does the supply. As resource capability changes, there needs to be a constant evaluation of how this change can be reconfigured. Regarding resource capability to support the change, it is not only about human resources. Non-human resources also need to be considered part of the change model. This would include technology tools.

Change management is a structured approach to transitioning organizations, individuals, and teams from a current state to a desired future state. However, when "change management" became popular in the business community, everyone was slightly confused.

For example, in the information technology industry, change management is related to standardizing business processes. The term change management is also used in systems engineering.

Describe the change management process, including how changes are requested and implemented. In HR, change management is about people's problems and how they respond to change. *Leaner* refers to CMP as related to change, including all the different perspectives on change management.

A common factor with all change management initiatives is resistance. People are resistant to change out of fear of losing something at a personal level. The *Leaner* Manufacturer may need to assure employees that the *Leaner* is a tool, not a replacement, for processes the employee currently controls.

Change management anxiety can also contribute to participants believing they do not have enough time to implement the change. Again, *Leaner* positions itself as a program that will help increase time and focus resources.

Leaner has several simple tools that can be implemented immediately to make things work faster.

There is a convincing argument that actual change agents should report to the executive leadership level. However, in some companies, change management is managed through the human resources department. Although this is not an ideal situation, if the company is invested in this idea, it would help the manufacturing project manager align with changing leaders in the human resources department. Therefore, change management is often covered in Human Resource Management programs.

Change management (professionals with essential manufacturing project management) backgrounds or training have apparent advantages. Being well-versed in an improvement methodology, such as *Leaner*, is an even bigger plus. The CMP, who can present a business case with an undeniable return on investment, will be respected. Mastering Cost/Time Estimation, Analytical Thinking Processes, and Fact-Based Strategies Open Doors. Understanding and using statistics to communicate your thoughts raises the bar for necessary conversations such as:

- Legal compliance related to change management
- Developing a recruiting and retention strategy
- Performance management
- Job design
- Knowledge management
- Human resource information systems
- Strategic, operational, and administrative issues

The role of the CMP is typically divided into three main categories: strategic, operational, and administrative. In today's world, the CMP is expected to act confidently in all three categories. *Leaner* methodologies provide a framework for confidently capturing and presenting information in all three areas.

There are many other ways the CMP may benefit from using *Leaner* concepts. First, the Leading Six Sigma methodology provides a solid path to implementing new manufacturing projects. Second, learning to use the statistical information promoted in Six Sigma allows the CMP to be better and more informed.

Third, studying Six Sigma principles improves communication with manufacturing project managers, technical staff, and executive management.

The CMPs, who are also responsible for health and safety issues, will benefit from applying the Six Sigma methodology to high-risk environments. Six Sigma forces the practitioner to study the existing system. This may lead to identifying potential dangers. Whereas many safety programs focus on satisfying lengthy compliance agendas, they do not position employees to consider future impacts or identify risk issues.

One area where change management professionals can make significant contributions is metrics design. It would be best if you distinguished which metrics represent the true added value of your organization. Measuring for the sake of measuring is time-consuming and wasteful.

Understanding departmental and organizational strategic initiatives is essential to deciding which metrics to use. Familiarity with strategic initiatives is critical to many HR processes, such as performance reviews, job descriptions, and employee orientation. All Lean models rely on sound measurement systems, and CMP may already compile valuable data for these metrics.

Many CMPs understand internal benchmarking. An internal benchmark compares one process within an organization to another, and the two operations do not have to be the same.

Exactly, but they should be similar. Knowledge of this process is beneficial for Six Sigma manufacturing project teams. Identifying the quality critical factor (CTQ) is not much different than placing the items necessary for employee satisfaction. Another popular *Leaner* model, SIPOC, Supplier, Input, Process, Output, Customer (SIPOC), discussed earlier in this book, is not a significant challenge for CMP. However, for the program to be successful, the recruitment process should identify these areas.

CMP is also good at documenting the best practices. This is another area where understanding the process is as important as understanding the subject. Even in structured *Leaner* manufacturing projects, the best internal practices are often not identified. This is usually because there is no way to communicate best practices. To be successful, organizations must implement processes that encourage and reward the sharing of ideas.

Scorecarding is one area in which CMP has design experience. Scorecards are recognized tools for tracking business success. A successful business scorecard balances long-term and short-term goals, financial and non-financial, and internal and external metrics.

Implementing a scorecard system requires translating vision into operational or financial goals. A CMP may not have company-wide experience, but most of his CMPs can do this type of measurement for their department.

This experience is helpful for manufacturing projects using scorecard systems, as it requires a commitment to a vision, process, and communication plan that can be shared with staff. The same abilities that enable manufacturing project managers to improve quality and results can be counterproductive without the necessary social skills.

Initially, obtaining the training necessary for a successful CMP may be difficult. You may not have.

The budget or desire to formally train your administrative staff. Other areas suitable for his CMP-seeking *Leaner* experience include developing *Leaner* retention strategies and writing job descriptions. Developing rewards and recognition programs is encouraged at *Leaner* but rarely has a severe process owner.

CMP should find a way to get involved. Our proactive approach is recognized and appreciated. Although formal education and training are required for a learner, hands-on experience makes concepts easier to understand.

The adoption of Learners in an organization is a major change that will have a significant impact on a wide range of stakeholders. Managers and employees at various levels of the organization are being asked.

To embrace the new behavior. Those leading other initiatives may see Lean as a source of competition for resources, executive attention, and organizational power. There can be confusion about how learners fit into the many ongoing organizational programs such as CMMI and ISO. It does not work without a plan.

Most leaders would say that people are their most valuable resource. It makes sense for change management groups to adopt and apply quality initiatives and continuous.

CMP helps you decide which features to measure and which metrics.

CMP is easy to use, allowing us to apply realistic benchmarks and provide training on how to create actionable scorecards. It helps reduce the uncertainty and anxiety associated with leaners and is a valuable resource for *Leaner* teams.

A CMP interested in researching and applying Six Sigma methodologies is an asset to any organization. You can improve your department's processes, act as a role model, and participate in larger company manufacturing projects.

Change Management Issues in Manufacturing

Change can be challenging in any industry, significantly if that change disrupts business as usual or your employees' daily routine. As manufacturing professionals, it is our job to plan and execute changes to be successful while ensuring that our employees and customers remain happy, productive, and satisfied with our products and services. Here are seven common types of change management issues in manufacturing, along with the strategies you can use to handle each of them successfully.

Here are the seven most common ones and how to oversee them:

- Employee Training
- Implementing New Technology
- Process Changes
- Changes in Leadership

Employee Training

One of the most common change management issues is that employees are not adequately trained for the changes. This problem can be a complex problem to solve, but there are some tips. You should have training sessions before and after the change and provide all necessary training materials as soon as possible, including videos or tutorials. Additionally, managers need to communicate with their employees about what will happen during the transition period to avoid surprises.

When all the changes take effect, managers should then give feedback to their employees on how well they did during the change process. Finally, tracking how healthy employees are performing after the transition is essential to identify any recent problems or challenges and address them as quickly as possible.

Implementing New Technology

The *Leaner* manufacturing professional needs to be aware of the following common fears when implementing modern technology.

- Lack of familiarity with modern technology.
- Fear that modern technology will make work more complex and less attractive.

- Belief that the old way was better and the new way was too complicated.
- The increased time it will take to learn how to use modern technology.
- Lack of communication about the changes.

By understanding the specific fears (listed above), the *Leaner* Manufacturing Professional can build a strategy and communication plan based on the audience. This would include education level, experience, locus of control, and environmental factors.

Process Changes

A process by which a product or service is manufactured, packaged, delivered, distributed, or used can undergo positive or negative changes.

In many cases, organizational change is implemented incorrectly because it is viewed purely as an implementation of a new process.

How can we get people to accept change? *Leaner* Leaders should encourage employees to embrace the new, whether the changes are big or small. Talk about how people feel. Discuss the employees' concerns and feelings and ask "what if" questions.

Changes in Leadership

Leadership changes can occur for assorted reasons, including changes in company ownership, restructuring, or layoffs. These changes may affect the company culture and create challenges for employees.

The biggest challenge for the *Leaner* Manufacturing professional is communication disruption. When no one is designated as the leader, employees might not know what they are supposed to do or to whom they need to report. Uncertainty about decisions – With so many different people making decisions, it is hard to stay on top of all the various opinions. It is also difficult to predict how long each decision will take.

Also relevant is limited time management skills – Employees do not have enough time to focus on manufacturing projects with competing deadlines because they are dealing with several priorities at once.

It is important to manage Poor delegation skills. Some managers find it difficult to delegate tasks due to a lack of confidence in their team members' abilities.

When leadership changes, there could be unrealistic expectations. Managers might push themselves too hard and expect unrealistic results from their employees if they have not been trained well enough in delegation, time management, etc.

Work overload is also a possibility. When employees accept their new position, they may work more hours than they originally agreed upon, which can cause fatigue and stress.

There could be a lack of clarity – Employees might feel like they do not understand the direction.

What is the company doing, and what work should they do?

The *Leaner* Manufacturing manager may have to deal with employee absenteeism – Employees.

They feel their needs are not being met anymore, so they seek other opportunities elsewhere.

Resistance to change – One common issue faced by companies undergoing a leadership change is resistance from those who were used to following the previous boss's orders and want things done their way now.

Other changes that need to be addressed are Organizational Changes, Customer Changes, and Economic Changes.

The person in charge of the manufacturing process can increase their success by paying close attention to changing management. Managing change will make any Lean Manufacturing Process *Leaner.*

Beliefs and values evolve with a company's history. They are not quickly abandoned. Change that does not address or respect these values and beliefs will not be successful. To function in a change management role, it is essential to do the following:

- Stay current on the organization's mission, policies, and plans.
- Be prepared to communicate the mission, policies, and plans.
- Act as a buffer between executive management and team members during stressful mergers, layoffs, or changes in direction.

Although change is an inevitable reality, framed correctly, it is possible to return to the past for inspiration. A mistake in essential change management is to entirely discount the old way of doing things. This is not necessary and can be counterproductive.

Team members cannot be considered malleable when it comes to changing management initiatives. Overseeing people like a sculptor molding clay into various forms is impossible. Even when the change is a positive one, there is pushback and resistance.

A surprising number of employee relations issues can be directly traced back to how healthy change was presented or managed within a group. There was a time when change was temporary and always followed by a more extended period of stability. Now, change is continual and does not allow team members time to regroup and accept the change before even a new change is imposed.

How an employee approaches the thinking process can determine how well that employee will adapt to change. Thinking freely, thinking analytically, and thinking holistically are all indicators that the employee can adjust quickly.

Individual skills and competencies can position individuals to accept change better than others as well, and these include the following:

Technical ability
Understanding project methodology Ability to create solutions
Capability of forming partnerships

Without natural competencies and skills, individuals exposed to change may benefit by learning about and implementing systems thinking. This thinking involves five effortless steps:

- Starting the problem
- Telling the story
- Identifying the key variables
- Visualizing the problem
- Creating loops

Stating the problem is the first step in every methodology. In areas dealing with change, it is crucial. Having the employee put the problem in story form helps the employee identify the issue more closely. Variables are components in the story that may change over time. Variables may include things such as a change in management. Visualization of the story in graphic form sometimes helps detect the change or behavior necessary. Finally, looping takes the story and illustrates which factors influence other factors. There are two types of loops:

- Reinforcing
- Balancing

Reinforcing loops are self-fulfilling prophecies, either positive or negative. Balancing loops keep things in equilibrium.

A collaborative approach to change almost guarantees high participation, strong commitment, and the creation of a reasonable standard that may be measured for results.

Executive management mandates are often given to the human resources (HR) department.

This may include delivering training and composing communications related to change management. While this approach can go wrong, the worst culprit is hurried communication.

A PM is expected to participate actively in the change management process. This can be challenging for a PM because the executive leadership team may expect a metamorphic change to occur overnight. Some PMs have been tempted to use e-mail to hurry things along when a face-to-face meeting would be more appropriate.

A popular definition of managing change refers to making changes in a planned and systematic fashion. Rather than allowing change to occur naturally and often randomly, change management assumes that it is possible to introduce planned change and steer its development.

PMS must learn to embrace and demonstrate good change agent skills. Internal changes may be triggered by events outside the environmental changes that are out of the manager's control. Implementing new methods and systems in an ongoing organization takes patience.

It is helpful to think of managing change in the same light as basic problem-solving because it involves moving from one state to another, just as problem-solving moves from the problem state to the solved state.

As a PM, it is essential to understand the company's philosophy to positively impact change. A PM's objectives must be embraced and aligned with the company's philosophy. Some costly mistakes can be made by a PM who has not been involved in the change management process. Areas to watch include the following:

- No systematic plan
- Under-communicating the vision
- Declaring victory too soon

William Bridges has pioneered change management and transitions. His material is still used in universities today.

Bridges believe that change is situational, such as moving to a new home. On the other hand, he describes the transition as the psychological process people undergo as they internalize and come to terms with the details of the new situation. Bridges divide transition into three phases:

- Ending phase
- The neutral zone
- The new beginning

Bridges asserts that it is impossible to move to the neutral zone if managers do not let go (the ending phase). Moving into the neutral zone is necessary if managers want to reach the final stage, the new beginning. These stages have also been referred to as unfreezing, changing, and refreezing.

Bridges is considered the pioneer of change management theory. His concepts were expanded on.

And adopted by Wharton's Center for Leadership and Change Management. They include the following:

- Stimulate basic research and practical application in leadership and change.
- Foster an understanding of how to develop organizational leadership.
- Support the leadership development agendas of the Wharton School and University of Pennsylvania. This means consistently updating data created by qualified groups of scholars.

In more traditional textbooks, change management has three primary areas that should be examined: the actual task of managing change, the new body of knowledge that must be delivered, and how the change will impact professional practice.

Change management requires political, analytical, people, business, and system skills.

Organizations are social systems. Without people, there can be no organization. Guessing will not do. Change agents must learn to disassemble and reassemble components, considering the financial and political impact.

There is no single change strategy. When developing a change strategy, it is essential to consider that successful change is based on information communication. Redefining and reinterpreting existing norms and values and developing commitments to new ones are also necessary for success.

Too often, a PM communicates the plan by simply reporting or posting it. A PM must take the time, even on smaller projects, to learn from Lean and Agile and prioritize understanding the voice of the customer, the voice of the employee, the voice of the process, and the voice of the business when communicating the plan.

Chapter 7

Basic Manufacturing Principles

Manufacturing is creating or manufacturing goods using labor, machinery, tools, and chemical or biological processes or formulations. It is the essence of the secondary sector of the economy. The term can refer to various human activities, from crafts to high-tech. Still, it is most applied to industrial design, where primary sector raw materials are transformed into finished products on a large scale. Such goods are sold to other manufacturers to produce more complex products (airplanes, appliances, furniture, sporting goods, automobiles, etc.) or sold to end users and consumers through the tertiary industry – distributors (usually sold by wholesalers through wholesalers).

The retailer that sells them then sells them to individual customers).

A manufacturing facility is a set of processes and systems (and people, of course) designed to transform a specific, limited range of materials into high-value products. These three building blocks of materials, processes, and systems make up the subject of modern manufacturing. There is a strong interdependence between these factors. Manufacturers cannot do everything. Only certain things are allowed, and these things must be done correctly. Manufacturing capacity refers to the technical and physical limitations of the manufacturing company and its facilities.

This capacity can be identified as:

- Technical Thoughts
- physical product limitations
- production capacity

DOI: 10.4324/9780367775414-7

Technical throughput or (Technological Processing Capability)

The technical capacity of a factory or a company is the available volume.

In the manufacturing process, some do machining, roll steel billets into sheet metal, and build automobiles. A machine shop cannot roll steel, and a rolling shop cannot make cars.

The fundamental feature that sets these plants apart is the processes they can conduct.

Technical workability is closely related to material type. Specific manufacturing processes are suitable.

Sure, processes or groups of processes are suitable. Specific processes or groups are ideal for certain materials, while others are suitable for others. By specializing in a particular process or group of processes, the facility also specializes in a specific type of material. Technical capacity includes not only physical processes but also the knowledge of plant personnel in these process technologies.

Physical Product Limitations (Size and Weight)

Physical products impose the second aspect of manufacturability. In facilities with fixed processes, there are limits on the size and weight of products that can be accommodated. Large and heavy products are challenging to move. The facility must have a crane with the required lifting capacity to move these products. Belt conveyors, etc., can convey small parts and mass-produced products. Product size and weight limits also extend to the physical capacity of manufacturing facilities. Production machines come in assorted sizes. Larger machines must be used to process larger parts. Production and conveying technology must be planned for products in specific sizes and weight ranges.

Production Capacity

A third constraint on a facility's manufacturing capacity is how much it can produce in a given period (such as a month or year). This quantity limit is usually called the capacity or production capacity of the plant. It is the maximum production rate the plant can achieve under expected operating conditions. Operating conditions relate to the number of shifts per week, hours per shift, direct workforce in the factory, etc. These factors represent the inputs to the manufacturing plant. Given these inputs, how much output can the plant produce?

Factory capacity is usually measured in units of production. For example, a large amount of steel is produced annually in a steel mill, or a car is made in a final assembly plant. In these cases, the output is uniform.

The first manufacturing industry was born out of necessity, but the Industrial Revolution changed everything. Soon, factories were popping up nationwide, and manufacturing significantly contributed to economic growth. As technology advanced, so did production methods. Today's industrial machinery is faster and more dependable than ever, making it easier for manufacturers of all sizes to compete internationally. Manufacturers constantly look for ways to improve their processes and reduce costs; this focus on innovation has led to rapid changes in many industries.

Whether a first-time entrepreneur or an established entrepreneur is looking to establish a manufacturing business in the United States, three imperative goals must be met. Although each state has its own industrial legislation and regulations, there are still some basics to keep in mind.

These basics include:

- Technology
- Quality
- Customer Service

There are three types of manufacturing. Each type has a different focus:

1. **Product Manufacturers**: These companies produce a product for sale. They may also produce components and parts for other manufacturers.
2. **Component Manufacturers**: These companies manufacture components assembled into final products by another company or person.
3. **Service Provider Manufacturers**: This group of manufacturers provides a service instead of a physical product, such as healthcare providers that make medical devices.

Four basic principles of manufacturing have been established and well known for many years. These principles are just as valuable today as they were when they were first introduced.

The first is known as make-to-stock. In this principle, all products are manufactured before they are sold.

The second principle is called make-to-order. Here, products are manufactured after an order has been placed.

Next up is designer fashion. In this principle, a collection of items from a designer is created and displayed at one time before being sold on demand. It is important to note that there is no inventory of designer fashion.

Finally, there is assembly-to-order. In this process, customers purchase preassembled kits or packages that include all the materials needed to construct the finished product. Once assembled by the customer, these goods are ready for sale.

Emerging technologies provide new ways to grow job opportunities in advanced manufacturing industries such as the US manufacturing belt. Manufacturing provides critical material support for national infrastructure and defense.

On the other hand, most manufacturing processes can have significant social and environmental costs. For example, the costs of removing hazardous waste may outweigh the benefits of the products that generate it. Dangerous materials can expose workers to health risks. These costs are now well known, and efforts are underway to address them by improving efficiency, reducing waste, utilizing industrial symbiosis, and eliminating hazardous chemicals.

Negative manufacturing costs can also be legally addressed. Developed countries regulate manufacturing activities through labor and environmental laws. Manufacturers worldwide may be subject to regulations and ecological taxes to offset the environmental costs.

Of their manufacturing activities. Unions and artisan guilds have played a historic role in bargaining for workers' rights and wages. Ecological laws and health and safety regulations in developed countries may not be available in developing countries. Tort laws and product liability creates additional production costs.

This is a critical dynamic in the ongoing process of manufacturing migration to 'developing' economies, where production costs are significantly lower than those of 'developed' economies, which have taken place over the past decades.

The Importance of Quality Control

Quality control is essential in manufacturing because it helps ensure that products are consistently high quality. If you want your company to succeed, you need a quality control strategy throughout your operations.

In most industries, two quality control systems are used: inspection and testing.

Inspection involves visually checking a product for defects, while testing involves examining the product for flaws under laboratory conditions.

These two methods complement each other by providing more assurance that a product is safe and dependable.

Inspections require less time than testing, usually done after production before the product is shipped to customers.

Testing must happen before production starts because prototypes are made and tested for safety and efficiency.

Companies often design their assembly lines so that components are manufactured with as few errors as possible during construction to reduce the amount of inspection required on final products.

Characteristics of Mass Production

Mass production has several characteristics to include

- Standardization
- Production Flow
- Undefined Demand
- Division of Labor

Standardization

The goal of mass production is to produce enormous quantities of homogeneous products. An organization's products should all be the same, without look, feel, and variation. For this reason, upper management must work to develop strict quality control standards.

Production Flow

The manufacturing process must be efficient so employees can move from development to packaging without returning. Assembly line delays, material delays, and downtime due to employee error are very costly to most manufacturing facilities. Therefore, a systematic O&M management program is essential to ensure the smooth operation of all systems.

Undefined Demand

The primary purpose of mass production is to reduce production costs that cannot meet fluctuating customer demand. Ideally, when demand for a product drops, companies respond by temporarily lowering their selling prices. Assuming the company maintains low unit prices, it should be able to sell cheaper items and still be profitable.

Division of Labor

The manufacturer divides the production process into stages. For example, one machine fills cans with green beans, and another seals the cans. Similarly, one worker can oversee the automatic labeling of cans on a machine, while another worker can drive a forklift to bring the wrapped boxes to the production warehouse.

At its core, manufacturing is an uncomplicated process. We purchase raw materials.

Or components and process them into final products. However, to be successful, manufacturers must be able to cover the cost of producing their products, meet demand, and create.

Desirable products on the market.

Companies should focus on designing and manufacturing products compatible with their technological, physical product, and production capabilities.

The manufacturing industry has its own health and safety challenges, and the National Institute for Occupational Safety and Health has identified and provided intervention strategies related to workplace health and safety issues under the National Occupational Research Agenda, recognized as a priority industry sector.

Another area common to manufacturing is the attempt to streamline. Streamlining is improving an item, system, or process to make it simpler or more efficient. For example, streamlining customer billing means improving the documentation process and making steps faster, repeatable, and consistent.

One method to make manufacturing lean is to streamline the processes. There are ten recognized steps to do this include

- Analyzing the workflow
- Identify the areas of focus

- Prioritize tasks
- Document
- Automate
- Evaluate and Test the New Process
- Make reasonable changes

The benefits of streamlining are saving money, improving communication, and better documentation.

Chapter 8

Lean Manufacturing NOW – a Fast Primer/Refresher

Addressing Lean and Manufacturing separately is a beautiful place to start. First, manufacturing will be examined, then lean. Once the two are summarized, this chapter will move into Lean Manufacturing.

Manufacturing is the transformation of raw materials, particularly on a massive scale. A standard definition of manufacturing is manufacturing merchandise or wares using labor or machinery.

Here is a summary of the standard industry procedure for manufacturing:

- Ideation – produce the idea.
- Ask for help/Get Assistance.
- Create the product or prototype.
- Plan the manufacturing project.

To sum up, a manufacturing process transforms raw materials into finished goods, with quality and performance metrics – among other things – coordinated to provide consistency and efficiency.

It does not matter the size of a company or the complexity of a product line; each factory will know the best production process for them, which is determined by the manufactured product. However, it is possible to see a wide variety of production processes used by different companies or within the same company.

There are common manufacturing tools and metal fabrication to include:

DOI: 10.4324/9780367775414-8

- Broaching machines
- Drilling machines
- Gear shapers
- Grinding machines
- Hobbing machines
- Honing machines
- Lathes
- Milling machines
- Planner machines

Broaching Machines

A broaching machine, or broach, uses tall chisel points to use linear cutting and scraping motions to the given material. Broaches are usually accustomed to producing ellipsoid shapes out of holes that are antecedently punched within the metal. They additionally cut splines and keyways on gears and pulleys. Rotary broaches square measure a singular section of broaching machines, employed with a shaping machine to make a synchronal horizontal and vertical cutting motion.

Drilling Machines

Drilling machines, called drill presses, include a fixed drill mounted or fastened to a stand or worktable. Drill press area units are employed in many constant ways, such as hand-held and power drills; however, the stationary nature of drill presses needs less effort to attain correct drilling and is more stable. Factors like the angle of the drill spindle may be mounted and maintained to permit continual and consistent drilling. Trendy drilling machine varieties embrace pedestal, bench, and pillar drills.

Gear Shapers

A shaper consists of a rotating workpiece upon which the feasible object (in this case, metal) is placed – the result is the symmetrical and specific shape of the merchandise. Because the product rotates, numerous tools are used to cut, knurl, drill, or alter the metal. The friction of rotation provides an easy mechanism for delivering a regular result around the entire circumference of the Associate in Nursing object, creating lathes an honest alternative for merchandise that area unit radially symmetrical around the associate in the nursing axis of rotation.

Grinding Machines

The grinder's area unit has trendy machining tools that use the associate in the nursing abrasive wheel to make fine finishes or faint cuts. Betting on the grinder, the abrasive wheel or product is rapt from facet to facet to attain the specified end. Various grinders embrace belt grinders, bench grinders, cylindrical grinders, surface grinders, and jig grinders.

Hobbing Machines

A hobbing machine is comparable to a miller in that a rotating cutter performs the cutting action. However, hobbing machines permit the coinciding movement of each cutter and, therefore, the product being machined. This distinctive capability makes hobbing ideal for 3D machining.

Applications that need uniform tooth profiles. Gear cutting is one of the most typical applications for contemporary hobbing machines.

Honing Machines

Honing machines, conjointly called hones, consist primarily of one or a lot of rotating tips that, in formation, enlarge holes to a definite diameter and improve the surface end. Varieties of honing machines embrace hand-held, manual, and automatic. Merchandise factory-made with the assistance of honing embrace engine cylinders and air bearing spindles.

Whereas a hobbing machine cuts the external teeth of a gear, trendy gear shapers fabricate internal gear teeth. Because of the cutting gear, this can be accomplished by employing a mutual cutter with constant pitch. Trendy gear shapers allow enhanced exactitude using forward stroke engagement and backward stroke disengagement.

Lathes

Lathes vary in size, with the littlest hand-held versions used for jewelry and watchmaking.

Milling Machines

Like drilling machines, milling machines use a stabilized rotating cutter to system a chunk of metal; however, they permit greater versatility through appearing sideways cuts. Some contemporary milling machines have a cell cutter, and others have a cell desk that moves a desk-bound cutter to finish the favored completing effect. Typical forms of milling machines consist of hand milling machines, simple milling machines, frequent milling machines, and universal milling machines. All forms of milling machines are available in vertical and horizontal configurations.

Planner Machines

Planners are big-sized shaping machines that pass the metallic product instead of shifting the reducing mechanism. The result is like a milling machine, making planners best for shaping flat or lengthy surfaces. Modern milling machines are advanced to planners in maximum applications; however, planners are helpful, while extraordinarily big metallic additives require squaring off.

Additionally, basic manufacturing processes are used. These types of manufacturing processes include:

- Repetitive Manufacturing
- Discrete Manufacturing
- Job Shop Manufacturing
- Process Manufacturing (Continuous)
- Process Manufacturing (Batch)
- 3D Printing

Repetitive Manufacturing

Repetitive production continues to manufacture an equal product for a prolonged time. The product is usually assembled on a manufacturing line, wherein a chain of responsibilities is finished within the equal series with the aid of personnel and robots. The amount produced no longer has a terminal value, and then manufacturing stops. Instead, a certain amount is centered on manufacturing in every successive duration.

Discrete Manufacturing

Discrete production is an enterprise period for producing completed merchandise that might be awesome objects that can be counted, touched, or seen without difficulty. Discrete production entails components and structures like nuts and bolts, brackets, wires, assemblies, and character merchandise. Examples of merchandise crafted from discrete production include automobiles, furniture, airplanes, toys, smartphones, and protection structures. Theoretically, a discrete product may be damaged down on the quilt of its lifecycle, so its primary additives may be recycled.

Job Shop Manufacturing

Job stores are small production structures that oversee task manufacturing, custom/bespoke or semi-custom/bespoke production strategies, and small – to medium-length client orders or batch jobs. They typically pass directly to jobs (with one-off customers) while completing every task.

Job stores aggregate machines based on the abilities and technological strategies involved. Consequently, every store might also incorporate machines, which offer this manufacturing machine processing flexibility, seeing that jobs are not permanently restricted to an unmarried gadget.

Other styles of non-unusual place task stores are grinding, honing, jig-boring, tools production, and fabrication stores.

Process Manufacturing (Continuous)

Continuous production, occasionally called go-with-the-drift production, is one of the predominant production approaches. It is a way that a product is made nonstop, with no interruptions or pauses among batches. Continuous manufacturing is historically used to produce high-quantity merchandise that does not require customization, inclusive of paper and pulp production, petrochemicals, or oil and refineries, and for merchandise that might be eating up on a persistent foundation inclusive of power manufacturing or water remedy plants.

Process Manufacturing (Batch)

Batch manufacturing is a way wherein a set of equal merchandise is produced simultaneously (in place of one at a time). The producer needs to determine how massive the batch may be and how regularly those batches may be made. Each batch goes together via separate levels of production.

3D Printing

3D printing, also known as additive manufacturing, is a new manufacturing process. 3D printing uses a digital model to create a three-dimensional product layer by layer from materials such as metal and plastic.

This manufacturing process has a higher initial cost but offers savings in other areas. Companies using 3D printing save financial capital, raw materials, and various forms of waste. 3D-printed products include buildings, musical instruments, shoes, firearms, prosthetics, and medical devices.

Learn

Lean is a mistake-proofing program that believes prevention is always better than detection and is a primary belief in securing process improvement. It concentrates on reducing or eliminating waste. Lean, in all its forms, is committed to doing things faster than normal process improvement methodologies.

These Lean Tools also apply to Lean Manufacturing as well as *Leaner* Manufacturing. They include:

- Fishbone
- Flowcharting
- Check Sheets
- Histogram/Frequency Diagram
- Pareto Chart
- Scatter Diagrams
- Control Charts

Additional Lean Tools also prevalent in Lean Manufacturing and are fully embraced by *Leaner* Manufacturing include:

Additional Charts and Graphs that are commonly used in Lean include:

- Value Stream Mapping
- Gantt
- PERT
- Swim Lane Charts
- Spaghetti Diagrams
- Tim Woods or the Eight Areas of Waste
- SWOT Analysis
- FMEA Thinking Process

Lean Manufacturing

Lean Manufacturing is a method that specializes in disposing of waste and maximizing performance. In lean production, all sports and procedures are scrutinized to become aware of any regions where waste is generated. Once identified, those regions are streamlined or removed to enhance overall performance. One of the essential ideas of lean production is to go with the drift of statistics and materials. In this method, there is no preventing or beginning inside the manufacturing process – the entirety flows quickly and constantly from one step to the next. This allows waste disposal and maximizes performance by ensuring there are no bottlenecks or delays within the manufacturing process.

There are five basic principles:

- Value
- Value Stream
- Flow
- Pull
- Perfection

Value

Knowing that any company's chief value is providing value to customers is essential to making the process as productive as possible. So, in other words, support and define business values by understanding them. This will allow the ability to work on vital tasks. It provides a competitive edge and quick delivery. It is a way of leveraging and differentiating the company from its competitors.

Competitive value can be determined in diverse ways, including direct conversation and cloud-based software that analyzes the data.

Value Stream

The Value Stream Map (VSM) includes all the business actions that convert raw materials into finished products.

The Value stream tracks the processes from raw material order creation to product delivery and maintenance. The idea here is to understand the factors and processes contributing to making the product valuable to customers. This often leads to discovering unnecessary steps in the process. In Lean Manufacturing, waste elimination plays a crucial role. Value stream diagrams help you identify and reduce waste.

The goal of VSM is to have a faster delivery time.

Flow

Once value is defined, value streams are mapped, and waste is minimized, only process steps that add value to the product remain. At this stage, Lean Manufacturing calls for streamlining the processes – this could involve breaking down each process into smaller steps, reconfiguring the production site, eliminating bottlenecks, or increasing production efficiency. The value stream is established with total site control and built-in end-to-end traceability.

Pull

Excess inventory leads to overhead costs of production. Examples of overhead include:

- Storage cost
- Loss of inventory
- Overspending on marketing products

Perfection

Lean manufacturing tools can help professionals improve their business.

And increase company profits. These tools help businesses save money by improving productivity and eliminating as much waste as possible. One of the Lean objectives is to strive for perfection. Striving for perfection does not always mean being "perfect" but pursuing perfection.

Most experts agree that these are the top tools used in Lean Manufacturing:

- 5S
- Bottleneck Analysis
- Cellular Manufacturing
- Gemba
- Heijunka
- Jidoka
- Just in Time (JIT)
- Kaizen
- Kanban
- Muda
- Overall Effectiveness (OEE)
- Poka-Yoke
- Total Productive Maintenance
- TQM
- Value stream mapping
- Visual Factory

A brief explanation of these tools follows. A more complete explanation is available in the Lean Manufacturing Terms at the end of this book.

5S is a set of organizational practices that create an environment conducive to efficiency, effectiveness, economy, and quality. Although developed in Japan by Toyota Motor Corporation, 5S can be used by any type or size organization. This method involves keeping areas clean, orderly, and well-organized with everything in its place; as a result, employees can find what they need when needed. They can work more productively because their tasks are simplified, and they do not have to waste time looking for tools or information. If the workplace becomes cluttered, this results in increased stress levels due to frustration and decreased productivity. 5S is a five-step approach to creating a more organized and productive workspace: Organize, Straighten, Shine, Standardize, and Maintain. 5S serves as the foundation for implementing more advanced.

Lean manufacturing tools.

A "**bottleneck**" is a work step that cannot meet production quotas even at total capacity, delaying or shutting down operations. The concept also applies to management and logistics.

Cellular **manufacturing** is a lean method for manufacturing related products using cells or groups of team members, workstations, or devices to support operations by eliminating unnecessary setup and costs.

It is needed between activities.

Gemba is a Japanese term meaning "real place." When used to improve the workplace or increase efficiency, Gemba is where the real work is done. For example, Gemba is where machines and other products are assembled in a factory.

The simple **heijunka** principle allows businesses to respond to changes in customer needs by establishing a standard or tiered workflow.

Jidoka is a principle implemented in lean manufacturing. In this system, machines automatically stop working when an abnormal condition is detected, and the operator tries to correct the error to prevent the problem from recurring.

Kaizen (continuous improvement) is a strategy in which employees at all company levels actively work together to achieve incremental and continuous improvements in the production process. In a sense, it brings together the collective talents within a company to create a powerful innovation engine.

Kanban (Japanese notation) is an inventory control system used in just-in-time (JIT) manufacturing to track production and order new materials. Kanban was developed by Taiichi Ohno, an industrial engineer at Toyota, and uses visual cues to trigger the action needed for a smooth process.

Lean manufacturing identifies three wastes that need to be reduced or eliminated: **Muda** (no value added), Mura (inequality), and Muri (overload).

Overall Efficiency (OEE) is the gold standard for measuring manufacturing productivity. Simply put, it determines what percentage.

Production time is efficient. A 100% OEE score means you only make good parts as fast as possible, with no downtime.

Poka-yoke is a lean production tool that refers to a process's "error override" or "error protection." In Japanese, poka-yoke is translated as "avoid" ("yoke") or "mistake" ("poka"). When a defect reaches the customer, it becomes a defect.

Total Productive Maintenance is a comprehensive approach to maintenance that strives for perfect production: No problems. Please do not make it short.

Total Quality Management (TQM) is the ongoing process of detecting and reducing or eliminating manufacturing defects, streamlining supply chain management, improving customer experience, and ensuring that employees are aware of the training.

Value Stream Mapping is a technique – evolved from Lean Manufacturing – that organizations use to create a visual guide to all the components needed to deliver a product or service to analyze and optimize the entire process. vision factory

Factory visualization is important in displaying valuable information, such as graphics and data, in the lean manufacturing workplace.

The Seven Tools of Quality in Manufacturing

When making decisions about improving the quality of your manufactured products, you need to consider every aspect of the process and how it will affect the way your product performs over time.

Quality in manufacturing does not end with the final inspection and stamp of approval – it is an ongoing process that involves design, construction, and much more! How do you know what matters when it comes to quality in manufacturing? Here are seven tools of quality in manufacturing that should be on every business owner's list. This chapter includes information about the graphs and the industry standard or accepted definitions.

Gathering information and using information regarding these tools can make your Lean Manufacturing process *Leaner*.

The tools discussed in this chapter are:

- Histograms
- Check Sheets
- Control Charts
- Cause-and-Effect Diagrams
- Scatter Diagrams
- Flowcharts
- Pareto Chart

Histograms

A histogram is a data visualization that displays the distribution of quantitative variables. It shows the frequency or probability density for values on the *x*-axis and frequencies or relative frequencies for values on the *y*-axis. Histograms can be used as quality control tools for potential quality issues.

Most experts agree with this definition: A histogram is a quality control tool that graphically displays a data set. More specifically, a histogram is a type of Bar Chart that graphs the frequency of occurrence of continuous data and will aid you in analyzing your data (Figure 8.1).

[Objective of Data Collection]

Name of operator: Sheet Number:

Date:

Place:

Reason	Frequency							
	Day 1	Day 2	Day 3	Day 4	Day 5	Day 6	Day 7	Total
[Issue 1]	×		×		×			3
[Issue 2]		×						1
[Issue 3]				×		×		2
[Issue 4]		×	×					2
[Issue 5]	×			×	×			3
[Issue 6]			×					1
[Issue 7]	×				×		×	3
[Issue 8]		×		×		×		3
[Issue 9]	×	×				×		3
[Issue 10]			×		×		×	3
Total	4	4	4	3	4	3	2	

Figure 8.1 Histogram.

For example, a sudden spike in defective products could indicate an issue with process controls. A histogram should also be part of any process map to help identify areas for improvement. Data visualization techniques like this are essential because they provide a clear picture of what is happening with the product. They allow you to quickly see what needs to be changed to improve your production methods and achieve better outcomes.

Check Sheets

Quality Control can be improved by using a series of check sheets. These are used to monitor the production process and provide documentation on how quality is being monitored. Check sheets are one method of quality control, but there are other methods as well. These include statistical process control, sampling inspection, and process capability analysis.

Most experts agree that A check sheet is a structured, prepared form for collecting and analyzing data. It is a generic data collection and analysis tool that can be adapted for various purposes and is considered one of the seven essential quality tools (Figure 8.2).

Control Charts

A control chart represents the variation between a process's meaning and upper and lower control limits. Control charts monitor processes that produce outputs with constant variances, such as those with no underlying

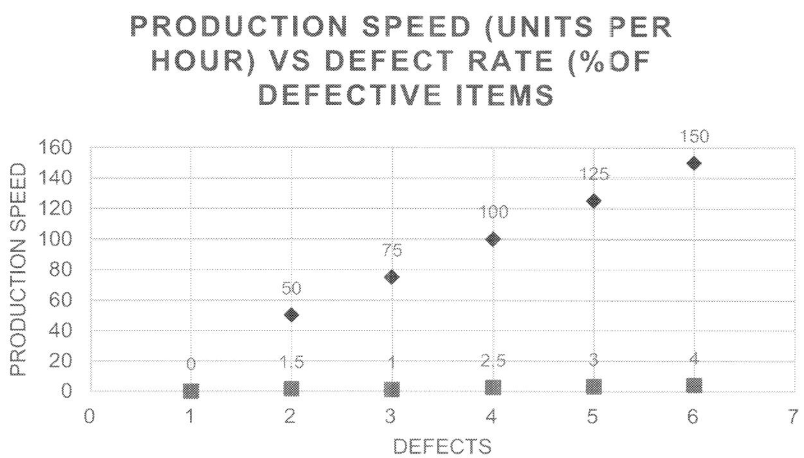

Figure 8.2 Check sheet.

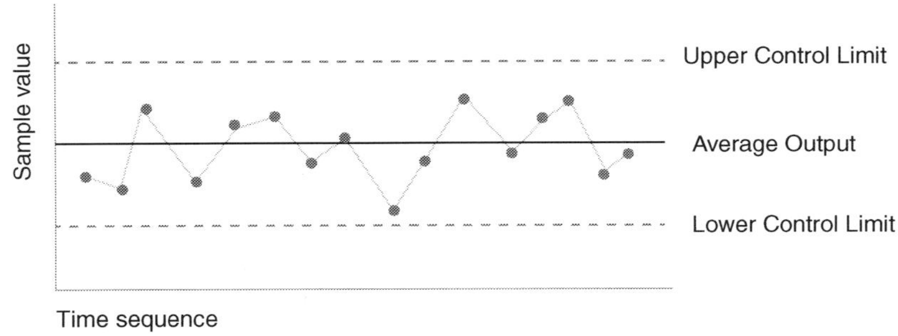

Figure 8.3 Manufacturing control chart.

trends or seasonality. A process can be monitored by taking samples at regular intervals, which should be spaced far enough apart so that they do not overlap. The data points are plotted on the chart according to their time order, not their date order. If a point falls outside the control limits, it signals an outlier.

Most experts agree that a control chart will help monitor, control, and improve process performance over time by studying variation and its source. It helps distinguish causes of variation from everyday causes and is a guide to local or management action (Figure 8.3).

Another measure of variability called the range includes high and low values: the difference between the highest and lowest values observed. Range tends to increase when there are more outliers, while Cpk is less sensitive to outliers because it compares average (mean) values to control limits instead of all values.

Cause-and-Effect Diagrams

Quality is the main factor that differentiates one product from another. To achieve high quality, it is essential to identify the seven tools of quality. These are: (1) Process Control, (2) Selection and Maintenance, (3) Statistical Process Control, (4) Design of Experiments, (5) Measurement System Analysis, (6) Procedure Writing, and (7) Standardization. Once these tools are identified, each must be executed precisely for an improved product (Figure 8.4).

Most experts agree that the fishbone or cause-and-effect diagram identifies probable causes for an effect or problem. It can be used to structure a brainstorming session. It immediately sorts ideas into functional categories. A fishbone diagram is a visual way to look at cause and effect. It is a more

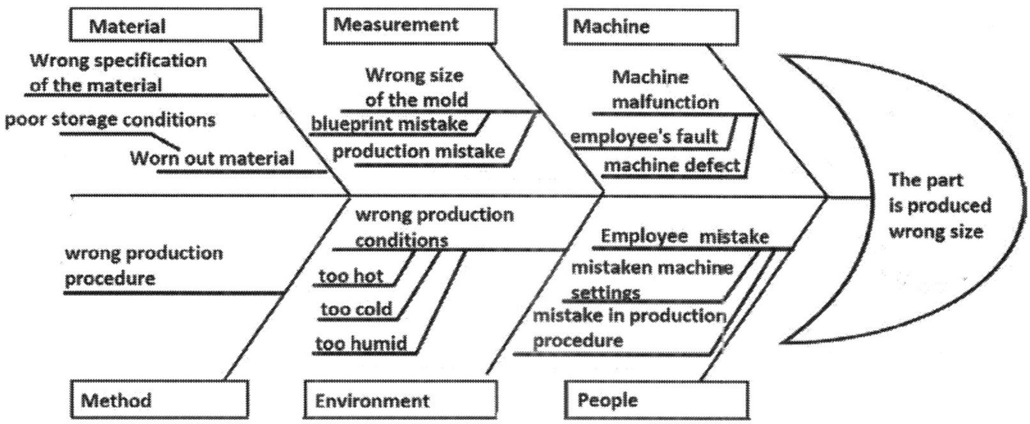

Figure 8.4 Cause-and-effect diagram.

structured approach than other tools available for brainstorming the causes of a problem (e.g., the Five Whys tool). The problem or effect is displayed at the head or mouth of the fish.

Scatter Diagrams

A scatter diagram is a powerful tool for understanding two variables' relationships. As with all other tools, its effectiveness depends on how it is used. Here are three best practices for using a scatter diagram:

1. Know what you are trying to show.
2. Pick an excellent variable representing your independent variable (*y*-axis).
3. Pick an excellent variable representing your dependent variable (*x*-axis) (Figure 8.5).

Most experts agree that the scatter diagram graphs pairs of numerical data, with one variable on each axis, to look for a relationship between them. This type of graph is also used to show correlations.

Flowcharts

Flowcharts in manufacturing may be used for:

1. Flowchart for incoming inspection
2. Flowchart for outgoing inspection

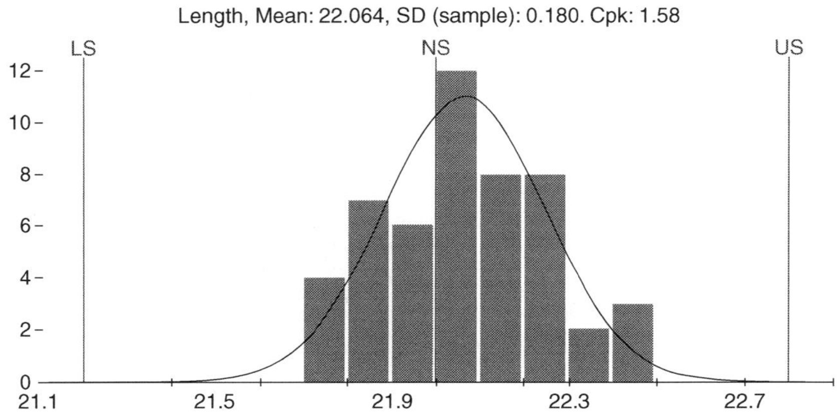

Figure 8.5 Scatter diagram.

3. Flowchart for design verification
4. Flowchart for process validation
5. Operation profile flow diagram (Figure 8.6)

Most experts agree that Flowcharts are the most popular among the seven quality tools. This tool is used to visualize the sequence of steps in a process, event, workflow, system, etc.

A flowchart shows the process as a whole and highlights the relationship between steps and the process boundaries (start and end).

Pareto Chart

A Pareto diagram illustrates a data set's distribution or relative frequency and size, revealing the essential contributions to those values. They are often used for quality control or for identifying the root causes when problems recur. This type of diagram is only one tool among seven for quality in manufacturing; it has limitations and should not be used alone but as part of an overall strategy (Figure 8.7).

The Pareto Chart can be presented graphically as either a bar graph, which displays bars stacked on each other in descending order of magnitude, or a line graph, which shows points plotted on an x–y axis with frequencies increasing toward the right (see example). Other tools, such as Check Sheets, can provide information about the criticality and urgency of defects found during quality inspections.

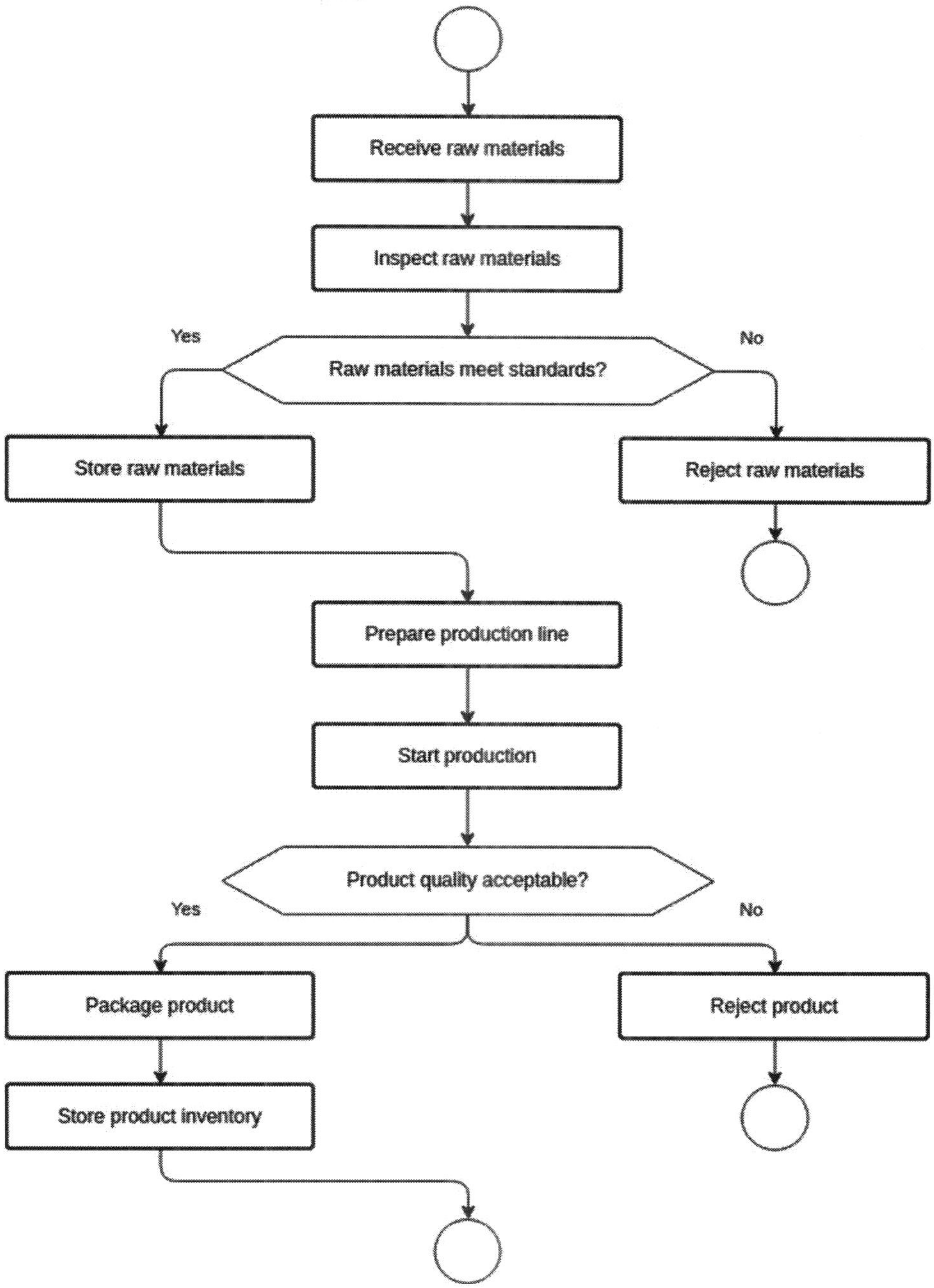

Figure 8.6 Flowchart.

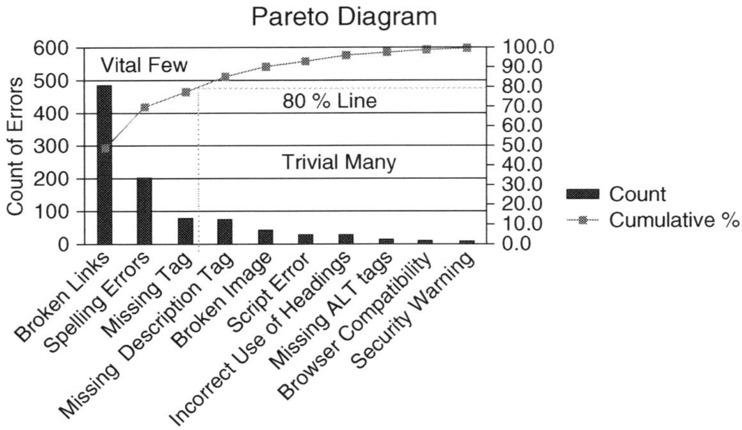

Figure 8.7 Pareto diagram.

Focused on improving productivity and eliminating unnecessary waste, Lean Manufacturing is ideal for companies looking to stay ahead of industry developments, such as the introduction of modern technology and changes in the workforce.

Chapter 9

Leaner Manufacturing™ for the Future

To make Lean Manufacturing leaner, it is crucial to understand basic manufacturing principles and concepts, as well as have a strong grasp of Change Management and Leadership. It is also necessary to embrace the seven tools of quality and have a complete understanding of the type of manufacturing being used.

One way to embrace Lean Manufacturing and make it *Leaner* is to train employees. The system explained in the next chapter will help employees connect the dots and understand the concepts faster.

Next, this chapter will discuss *Leaner* Manufacturing™ as this work sees it in the future and how to make it even *Leaner.*

Lean manufacturing is not a trend or a fad; it is an ethos that will prove its worth repeatedly in the next decade. *Leaner* Manufacturing™ is designed to simplify the key concepts and gain momentum. Think of *Leaner* as simply the next generation of Lean Manufacturing.

Remember that the benefits of Lean Manufacturing have already been accepted. Lean manufacturing also minimizes waste and maximizes productivity, which is essential for any company, especially companies looking to survive during tough economic times.

Companies should research lean manufacturing more before deciding not to use it because they feel they do not have time or money. There are many things companies can do to help with implementation, such as being open-minded, hiring good employees, and starting small when possible.

DOI: 10.4324/9780367775414-9

Once these *Leaner* Manufacturing™ strategies are implemented, companies will see a significant increase in profits and customer satisfaction.

One of the biggest things about *Leaner* Manufacturing™ that improves Lean Manufacturing is that it promotes keeping up with technology. When Lean Manufacturing first became popular, the software and technology were not as powerful as they are now. When certain technologies were adopted in Lean Manufacturing, they were not used to maximize the process but rather as templates to keep the older technology going.

Today's software offers intelligent manufacturing and options that allow tracking key process indicators (KPIs) and optimizing efficiencies in data exchange. Kanban software has been improved and made more cost-effective. Most competitive vendors have a try-it-before-you-buy-it option. Software and Artificial Intelligence are constantly evolving. Keeping up with those developments will make Lean Manufacturing *Leaner*; these hybrid packages also offer many organizational and reporting tools to benefit the manufacturing process.

Leaner Manufacturing™ promotes an investment in Artificial Intelligence (AI).

While most people are aware of artificial intelligence's (AI) impact on the world around them, many do not realize just how dramatically it will shape the future of manufacturing. Let us look at how AI will alter factory life as we know it today and explore how this innovative technology can create more efficient, effective, and safer workplaces.

AI can help manufacturers save time and money while streamlining their processes. With the rise of artificial intelligence, manufacturers have achieved this goal by using robots that can do work that humans would typically do.

For example, manufacturers have succeeded in using robots for quality control or other tasks where human error is prevalent or a human's attention span is limited. Robots can also take on repetitive, labor-intensive jobs so humans can do more creative tasks.

AI can improve manufacturing quality. Every industry is impacted by artificial intelligence. In manufacturing, AI can improve quality control and reduce defects. This reduces waste and saves manufacturers' costs. It also leads to higher productivity as humans are freed from mundane tasks to work on more meaningful manufacturing projects.

Today advances in automation and artificial intelligence mean that factories could become places where people can get back to making things. Robots could run the production lines so that humans can focus on designing new products or fixing broken ones. The human mind could be given back an opportunity to create instead of just assembling parts all day.

AI can analyze data, which can help manufacturers reduce waste. For example, if a part has a manufacturing defect or an incorrect quantity, an algorithm can pick it out more quickly than a human Furthermore, since machines do not need breaks or have emotions that interfere with their work, they are better at repetitive tasks and doing jobs that would otherwise accept too much time for humans.

AI can help improve customer satisfaction by providing a more customized product. It can use customers' preferences to suggest variations on products, providing a more personal experience. This is just one-way artificial intelligence can help manufacturers improve customer satisfaction.

Lean Manufacturing also addresses the challenges of Lean Manufacturing. The first challenge is establishing a new vision for lean manufacturing. It is not enough to keep going with what we are doing now because learning should be better. It should eliminate waste, not just reduce it. The second challenge is creating a new way of thinking about learning. We need more than just the tools; we also need a different culture and mindset. The third challenge is getting everyone on board with lean.

Establishing a New Vision

As the United States manufacturing sector continues to decline, it has become clear that drastic changes must be made to reverse this course and start bolstering growth once again. We need to establish new visions for manufacturing that will lead to change across multiple dimensions of industry, from both an individual perspective and a governmental level. If we want to ensure the continued growth of our country's manufacturing sector, we need to do so now, before it is too late.

Manufacturing has been changing rapidly in recent years, and it is essential to recognize how these changes can benefit the future of manufacturing worldwide. Recent studies have shown that manufacturing needs to undergo a dramatic shift if it wants to remain competitive in the global economy.

Manufacturing has traditionally been the backbone of the American economy, powering innovation and job creation. But, as technology advances, manufacturing jobs are disappearing. If we do not act now, our competitiveness in the global market will continue to slide. That is why it is time for a new vision for manufacturing.

While students are still required to take four years of math and science in high school, they are not required to take any vocational or technical classes. As a result, there is an extreme disconnect between what high schoolers think they will be doing when they grow up and reality. To bridge this gap, states should allow students pursuing technical careers to study those trades alongside traditional academic courses such as English and math. Doing so will enable them to enter the workforce with skills specific to their interests while having knowledge that can transfer into many different fields.

Moreover, these trade jobs often require much less formal training than other occupations, so these individuals would have far less debt when they graduate and enter the world. One way to encourage companies to invest more in manufacturing is to provide tax incentives for businesses that open domestic or overseas factories. If these two things were done, manufacturing could once again become one of America's most significant industries.

The last step would be to change our overall mindset about the field. The days of making fun of someone who works at a factory should end because, just like other types of work in America, factory jobs are respectable professions that give people opportunities and paychecks they might not get elsewhere.

Creating New Ways of Thinking about Lean

This book supports the idea that Lean would be more easily used and embraced if the concepts and methodologies were simplified. This author has been dedicated to that pursuit in prior books.

Lean manufacturing offers many great tools and techniques that can be used in any manufacturing or service industry. With Lean education, employees are taught tools and techniques that improve their process management, make jobs more enjoyable and efficient, and reduce waste and defects. However, this education must adopt new ways with each new generation of learners. Older techniques may no longer work or may not effectively reach every employee within the company to produce results.

Traditional education has problems with Lean. One is a simple lack of courses or knowledge at the secondary education level.

Education has become costly and time-consuming at the traditional college level. The average student spends $27,000 on tuition, books, and room and board in a four-year degree program. Many students are graduating with high debt levels and are still looking for jobs that match their education. This traditional path is not sustainable for today's world, where time is more valuable than money. The issue with conventional education has started a new movement toward more efficient and effective learning. This approach to lean education leverages existing technologies, tools, and resources to deliver meaningful learning experiences more efficiently.

One answer is more effective E-Learning on Lean Topics. Implementing an Intranet with Just-in-time manufacturing topics in Lean would be beneficial. This way, employees could look up issues related to Lean topics. Again, with AI advancements, the system can anticipate the right questions as it learns and grows with the company.

Other ways to make education more effective include:

- Gamification, which is the use of game mechanics in teaching. It has been shown that gamification can improve learning outcomes and engagement.
- Leveraging Social Media networks like Facebook and Twitter to deliver blended learning programs and offer access to online courses for learners who cannot attend live classes.
- Encouraging learners to collaborate using social media tools like Facebook, Twitter, and Google+.
- Offering personalization by having learners answer pre-work or workbook questions before accessing course content.
- Offering a choice between video, audio, or written instruction formats so all learners can engage according to their preferences.
- Taking advantage of mobile devices by offering mobile-friendly versions of e-learning courses.

Getting Everyone on Board with Lean

Buy-in for new and established employees begins with why buy-in is so necessary.

A buy-in demonstrates that employees understand the value they bring to the organization. This understanding is critical as companies and their employees face the challenges of growing their businesses.

Streamlining Manufacturing

Another critical topic for Lean Manufacturing *Leaner* is streamlining manufacturing processes.

Streamlining means reducing waste, improving quality, and leveraging available technology to enable all team members to work together. This can improve material and labor costs.

There are challenges in Lean manufacturing that can be remedied with *Leaner* Manufacturing™. Much of *Leaner* success is simply mindset. For example, if we stop thinking of manufacturing as the old, dirty industry on its way out, we see it as a new opportunity with endless potential. To make this happen, creating a new vision for manufacturing in the U.S. is necessary, and this starts by making changes in our education system.

In the past, training was primarily delivered in a traditional classroom setting with a long-term commitment. Today, companies are looking for more flexible ways of delivering lean education.

Instructor-led Online Courses – Classroom training is being replaced by instructor-led online courses, which provide greater flexibility and lower costs because it is not necessary to pay for a physical building or travel. However, one-way learning models do not have the same success as workshops taught in real-time by a live instructor. Virtual classrooms, where students interact with other students and the facilitator, are more successful than video production.

Employee approval positively impacts a company's transformation and growth efforts. However, employee buying outweighs the impact on a company's reputation and bottom line.

Looking at new ways to manage old problems, such as bottlenecks, can make a Lean Manufacturing Leader. Optimizing the process (streamlining) reduces the time it takes for employees to complete it without error. In addition, it reclaims valuable time and productivity that employees can use to work toward other goals and objectives.

Chapter 10

A Fully Developed Training Plan to Implement

Leaner Manufacturing™

One of the items to consider in the learning and development of *Leaner* Manufacturing™ employees is how information is presented. This fully developed training plan allows the student to embrace key learning topics and build on those concepts each day. Companies may elect to do this training in a boot camp style or take a day each week or each month.

Training in a regular Lean Manufacturing environment is often fragmented and on the job in reactive scenarios. Whereas this training may be done GEMBA style (on the manufacturing floor), students who understand the concept before applying the core tools have a great advantage.

Leaner Manufacturing™ Course Description

Day One	*Day Two*	*Day Three*	*Day Four*	*Day Five*
Lean Intro	Value Stream	5S	SPC	DOE
Terminology	Lean Manufacturing projects	FMEA	DMAIC	SMED
Components	Lean Tools	SIPOC	WIP	TOP

(*Continued*)

DOI: 10.4324/9780367775414-10

(Continued)

Day One	*Day Two*	*Day Three*	*Day Four*	*Day Five*
Lean Thinking	Application	Project Management	VS Mapping	Add'l Tools
Waste Mgmt.	Error Proofing	Pull System	Standardization	Group Exercise
Practice	Practice	Practice	Practice	Practice
Group Exercise	Group Exercise	Group Exercise	Group Exercise	Review
Quick Quiz	Quick Quiz	Quick Quiz	Quick Quiz	Final Quiz

Day One

Pre-Test

On the first day of class, students are given a multiple-choice test. This test will be compared against a test taken at the end of the class to measure progress. In the Lean-only classes, this is not a graded test but rather a way for the students to measure their progress and provide information on areas of self-study.

Introduction to Lean

Lean production is the value of a product based solely on the customer's point of view. Products must meet customer needs, both at a specified time and price. The thousands of mundane things needed to deliver a product are of little interest to the customer. To see the value from the eyes of the customer, most companies need to analyze all their business processes comprehensively. Identifying the value in Lean Production means understanding all the activities required to produce a particular product and optimizing the entire process. This section discusses the basics of Lean Manufacturing.

Lean Terminology

Lean has a language of its own. We have separated lean terms into six subdivisions: General Terms, Workflow, Workplace Organization, Workplace Simplification, Process Improvement, and Measures.

Lean Components

This section gives an overview of several topics explained in depth later in the course. The components include Value Stream Mapping, 5S, Quality, Mistake Proofing, Set-up Reduction, Total Productive Maintenance, Visual Workplace, and continuous improvement via the PDCA (Plan-Do-Check-Act) Model.

Waste Management

Lean methods typically target eight types of waste, each with a potential environmental impact, shown below. Interestingly, the "wastes" typically targeted by ecological management agencies, such as non-product output and raw material wastes, are not explicitly included in the list of manufacturing wastes that lean practitioners routinely target.

Practice, Group Exercises, Quiz

This section focuses on identifying various lean tools, models, and terminology needed to be successful in a Lean Manufacturing environment. Practice on PDCA, eight Areas of Waste, and essential fives

Day Two

Day One	*Day Two*	*Day Three*	*Day Four*	*Day Five*
Lean Intro	Value Stream	5S	SPC	DOE
Terminology	Lean Manufacturing projects	FMEA	DMAIC	SMED
Components	Lean Tools	SIPOC	WIP	TOP
Lean Thinking	Application	TMANUFACTURING PM	VS Mapping	Add'l
Waste Mgmt.	Error Proofing	Pull System	Standardization	Group Exercise
Practice	Practice	Practice	Practice	Practice

(Continued)

(*Continued*)

Day One	*Day Two*	*Day Three*	*Day Four*	*Day Five*
Group Exercise	Group Exercise	Group Exercise	Group Exercise	Review
Quick Quiz	Quick Quiz	Quick Quiz	Quick Quiz	Final Quiz

Value Stream Mapping

Value Stream Mapping is a technique for analyzing the flow of information currently required for a product or service to reach a consumer. It is often used in the environment to identify opportunities for improving delivery time. Value Stream Mapping is usually associated with production but is also used in coordination, supply chain management, services, software, and product development.

Lean Manufacturing Projects

Lean Manufacturing project Planning is the first step toward the timely completion of any manufacturing project. The process traces the flow of information. Lean manufacturing project management leads to more successful manufacturing projects. This includes satisfied customers and the manufacturing project team, as well as full-scope, on-time, and under-budget manufacturing projects. This training section also covers how to reduce pressure on manufacturing project team members and how to reduce unnecessary manufacturing project paperwork.

Lean Tools

This section gives an overview of several tools discussed in-depth later in the program, such as Conflict Resolution Diagrams, Future Reality Diagrams, Inventory Turnover Rate, JIT, Kaizen, Kanban, and Voice of the Customer, and Critical to Quality.

Application

This section applies Lean Tools to Lean Manufacturing projects. It is designed to enhance the previous two learning modules: Lean Manufacturing Projects and Lean Tools.

Error Proofing

Product Quality is the key to a successful business. Manufacturing processes, machines, human error, and many other factors contribute to product failure. This can be frustrating, as customers demand and expect zero defects. Quality is not free. Poka-Yoke (error proofing) is a technique to prevent product defects caused by human error. Defects have two states: They are about to happen (destruction) or already happened (detection).

Practice, Group Exercises, Quiz

This section focuses on using various lean tools and working with lean manufacturing projects and case studies. Error proofing and identifying suitable lean manufacturing projects are also part of these exercises.

Day Three

Day One	*Day Two*	*Day Three*	*Day Four*	*Day Five*
Lean Intro	Value Stream	5S	SPC	DOE
Terminology	Lean Manufacturing projects	FMEA	DMAIC	SMED
Components	Lean Tools	SIPOC	WIP	TOC
Lean Thinking	Application	Project Management	VS Mapping	Add'l
Waste Mgmt.	Error Proofing	Pull System	Standardize	Group Exercise
Practice	Practice	Practice	Practice	Practice
Group Exercise	Group Exercise	Group Exercise	Group Exercise	Review
Quick Quiz	Quick Quiz	Quick Quiz	Quick Quiz	Final Quiz

5S Model

The step in Lean requires a clean, streamlined factory floor so that everyone involved can focus better on production processes and material flow. 5S is a housekeeping methodology for the shop floor. There are five housekeeping rules for a lean environment, and they help expose waste.

FMEA

Failure Mode and Effects Analysis (FMEA) is a procedure for analyzing potential failure modes within a system for classification by severity or determination of the failure's effect on the system. It is widely used in manufacturing in various product life cycle phases. Failure causes any errors or defects in process, design, or items, especially ones that affect the customer and can be potential or actual. The effects analysis refers to studying the consequences of those failures.

SIPOC

A SIPOC diagram is a tool used by a team to identify all relevant elements of a process improvement manufacturing project before work begins. The SIPOC tool is handy when it is unclear: (1) who supplies inputs to the process, (2) what specifications are placed on the inputs, (3) who the actual customers of the process are, and (4) what the customer requires.

Pull System

Pull is a method of controlling the production flow through the factory based on a customer's demand. Pull Systems control the flow of resources in a production process by replacing only what has been consumed. They are customer order-driven production schedules based on actual demand and consumption rather than forecasting. Implementing pull systems can help you eliminate waste when handling, storing, and getting your product to the customer.

Practice, Group Exercises, Quiz

This section focuses on designing and examining fives, SIPOC, MANUFACTURING PM, and Pull System Models.

Day Four

Day One	*Day Two*	*Day Three*	*Day Four*	*Day Five*
Lean Intro	Value Stream	5S	SPC	DOE
Terminology	Lean Manufacturing projects	FMEA	DMAIC	SMED
Components	Lean Tools	SIPOC	WIP	TOC
Lean Thinking	Application	Project Management	VS Mapping	Additional Tools
Waste Mgmt.	Error Proofing	Pull System	Standardization	Group Exercise
Practice	Practice	Practice	Practice	Practice
Group Exercise	Group Exercise	Group Exercise	Group Exercise	Review
Quick Quiz	Quick Quiz	Quick Quiz	Quick Quiz	Final Quiz

SPC – Statistical Process Control

Statistical process control (SPC) involves using statistical methods to measure and analyze process variation. It is used for manufacturing processes. SPC aims to monitor product quality and maintain processes to fixed targets. SPC monitors the consistency of the processes used to manufacture a product as intended. The aim is to get and keep processes under control. No matter how good or bad the design is, SPC can ensure that the product being manufactured is designed as intended. A primary tool used for SPC is the control chart. A Control Chart is a graphical representation of certain descriptive statistics for specific quantitative measurements of the manufacturing process. The type of quality characteristic often classifies the types of charts that they are supposed to follow. There are quality control charts for variables and control charts for attributes.

DMAIC – Define, Measure, Analyze, Improve, and Control

The DMAIC model typically used in Six Sigma is rapidly becoming popular in Lean Manufacturing. This section provides an overview of this robust methodology. The DMAIC model has a series of five steps designed to provide a proper framework for successful manufacturing projects. Experts, including Six Sigma Black Belts, swear by this model for overall business success through quality control, problem management, and common sense. This section briefly overviews how the model is applied in a Lean Manufacturing environment.

WIP – Work in Progress

A core Lean Manufacturing philosophy is to reduce the number of things in WIP, which leads to efficiency and better quality. This section discusses tools and activities that can reduce WIP, which is work that has not been completed but has already incurred a capital investment from the company.

Value Stream Mapping

This section explains value stream mapping, a concept learned earlier in the program. Value stream mapping is a paper-and-pencil tool that helps you see and understand the flow of material and information as a product or service moves through the value stream. This tool may also now be computerized. An introduction to VSM symbols is included.

The first stage of any Lean improvement activity is identifying the product or service value the customer/end-user is willing to pay. Following this, the creation of value needs to be identified and arranged in such a way as to minimize the amount of waste created. Value Stream Mapping (VSM) and Analysis is a tool that analyzes the processes or activities of a manufacturing operation. These activities can then be divided into ones that do and do not add value. Mapping involves producing a diagram that describes the process. Value Stream Mapping can be carried out using pen and paper for a quick representation, or software can be purchased to aid in constructing maps and analyzing value streams.

Standardization

A key concept in Lean Manufacturing is standardization. This section discusses recognizing which processes should be standardized and calculating the time and resources saved by this process.

Practice, Group Exercises, Quiz

This section focuses on building value stream maps and working with Control Charts.

Day Five

Day One	*Day Two*	*Day Three*	*Day Four*	*Day Five*
Lean Intro	Value Stream	5S	SPC	DOE
Terminology	Lean Manufacturing projects	FMEA	DMAIC	SMED
Components	Lean Tools	SIPOC	WIP	TOC
Lean Thinking	Application	Project Management	VS Mapping	Add'l Tools
Waste Mgmt.	Error Proofing	Pull System	Standardize	Group Exercise
Practice	Practice	Practice	Practice	Practice
Group Exercise	Group Exercise	Group Exercise	Group Exercise	Review
Quick Quiz	Quick Quiz	Quick Quiz	Quick Quiz	Final Quiz

DOE-Design of Experiments

A Design of Experiment (DOE) is a structured, organized method for determining the relationship between factors (Xs) affecting a process and the output of that process (Y). Taguchi methods, developed by Dr. Genichi Taguchi, refer to techniques of quality engineering that embody both statistical process control (SPC) and new quality-related management techniques. Most of the attention and discussion on Taguchi methods has been focused on the statistical aspects of the procedure; the conceptual framework of a methodology for quality improvement and process robustness needs to be emphasized. The entire concept can be described in two fundamental ideas: (1) Quality should be measured by the deviation from a specified target value rather than by conformance to preset tolerance limits, and (2) Quality cannot be ensured through inspection and rework but must be built in through the appropriate design of the process and product.

SMED-Single Minute Exchange of Die

Single Minute Exchange of Die (SMED) is one of the many lean production methods for reducing waste in a manufacturing process. It provides a rapid and efficient way of converting a manufacturing process from running the current product to running the following product. This rapid changeover is the key to reducing production lot sizes and improving flow (Mura), a 'Lean' aim. It is also often referred to as Quick Changeover (QCO). Performing faster changeovers is essential in manufacturing or any process because it makes low-cost, flexible operations possible.

TOC-Theory of Constraints

Theory of Constraints (TOC) is a comprehensive management philosophy aimed at eventually achieving more of the goal of a system. If this system is a for-profit business, the goal is to earn more money now and in the future. TOC comprises two primary collections of work: (1) The five focusing steps and their application to operations, and (2) The thought processes and their application to manufacturing project management and human behavior. According to TOC, every organization has a critical restriction that limits system performance relative to its target. This section discusses how to recognize and control these limitations.

Practice, Group Exercises, Quiz

This section focuses on exercises and material related to DOE and TOC.

Review

An extensive review of the course is done with student study groups. FINAL ASSESSMENT

This is a final quiz, graded in class by the students to help each student identify areas they may want to study later. Students can look at the Pre-Test on the first day of class to see their progress.

Additional Tools

This section introduces additional tools and concepts that could be incorporated into this training:

Additional Charts and Graphs that are commonly used in Lean include:

- FMEA Thinking Process
- Gantt
- PERT
- Spaghetti Diagrams
- Swim Lane Charts
- SWOT Analysis
- Tim Woods or the Eight Areas of Waste
- Value Stream Mapping

Less commonly used tools/concepts that are recognized.

- *3 Ps*
  - People
  - Product
  - Process
- *6 Ms*
  - Machines
  - Methods
  - Materials
  - Measure
  - Mother Nature
  - Manpower
- *6 Ws*
  - What?
  - Were
  - Why?
  - Who?
  - When?
  - Which?

- *8 Ds – Short for the Eight Disciplines*
  - Establish the Team
  - Describe the Problem
  - Develop an Interim Containment Plan
  - Determine Root Cause
  - Choose Corrective Action
  - Implement Action
  - Preventing Recurrence
  - Recognize the Team

Here are a few helpful tips for any workshop setting:

- Keep the group focused. When someone is distracted, writing, or not paying attention, call their name and ask for their opinion (people tend to pay more attention when their name is mentioned).
- Engage everyone.
- Make sure everyone is comfortable.
- Use ice breakers when appropriate.
- Do not put too much information on the presentation slides.
- Stop and ask questions.

The key to a successful workshop is involving all stakeholders. If he stands and speaks for three hours, he will only give a presentation and not facilitate a workshop.

Everyone should participate!

This chapter includes a bonus for implementing *Leaner* Six Sigma® into your administrative practices. The term *Leaner* is the training methodology. This implementation may be used with manufacturing employees who are in operational and administrative roles.

This summarizes the advantages of combining Lean and Six Sigma methodologies followed by an implementation plan. However, a more in-depth understanding of Lean Six Sigma can be found in Appendix 3*: Lean Six Sigma in Manufacturing.*

What are the benefits of combining Lean and Six Sigma methodologies? What role do they play together, and how can both be used in the workplace?

These are frequent questions about whether a methodologist wants to use them at work or in business. By learning about the differences between the two and their specific benefits, the individual in training can decide whether it is worth implementing one or both in their workspace.

Lean and Six Sigma are two popular approaches to management and quality control, but what are the benefits of combining these methodologies? In this guide, we will outline some keyways these methodologies complement each other and some essential reasons to approach them as separate processes.

Lean, as promoted in this book, is a methodology that focuses on reducing waste in processes. This is done by finding ways to make the process more efficient, eliminating unnecessary steps, or simplifying procedures. The goal is to cut costs and deliver better customer service.

Motorola developed Six Sigma in 1986 as a statistical analysis-based production improvement technique. The methodology comprises five phases: reducing defects through each phase, analyzing, improving, and controlling.

Lean and Six Sigma are robust methodologies for creating a powerful hybrid process. This post will discuss the benefits of creating a combined method, why you should do it, and how to get started.

The combined methodology of Lean and Six Sigma is a perfect way to streamline operations and improve quality in your company. The two methods can be blended, but knowing their differences is essential before you start. Lean is a philosophy that focuses on eliminating waste in any system. It helps an organization focus on what it needs while removing anything that does not provide value.

The following information is a template for implementing this methodology for non-manufacturing employees, operations, and administrative staff to have a *Leaner* office.

As noted and explained in *Leaner* Six Sigma® by Francis & Taylor, 2019, it is interesting to consider that other process improvement methodologies need the buy-in or approval of the department head. In the case of ISO or CMM, the CEO must endorse the effort. One of the things that makes Lean Six Sigma different is that the tools and the methodology can be applied and used immediately by an individual employee. It can also be applied as a grassroots effort. Executing Lean Six Sigma basics requires no permission to get started. It is about the attitude and approach to projects.

Having said that, there are still some basic things the employee must embrace to make applying Lean Six Sigma to the workplace easier. The most important thing to remember is that Lean Six Sigma assumes basic project management is in place and understood. It is essential that the employee or the department has a high-level understanding of project management issues such as basic budgeting principles and return on investment.

Other areas of knowledge necessary to be successful in the implementation process are Risk Management, which includes an understanding of Risk

Analysis and Risk Mitigation, and basic people management, which includes topics such as leadership, role definition, and project selections that align with the available human resources.

- Basic project management activities include:
- Planning
- Assessing Risk
- Allocation of Resources
- Acquiring Human and Non-Human Resources
- Primary tools used in project management include:
- Project Charter
- Process Map
- Stakeholder's Analysis
- Responsibility and Accountability Document
- Work Breakdown Structure
- Gantt Chart
- PERT Chart

Should a Lean Six Sigma Black or Master Black belt be planned to lead a full-scale implementation of Lean Six Sigma that includes internal certification, there is no reason the internal program cannot use the same martial arts designations. Although an internal certification is less potent than one from an outside vendor, the advantage of internal certifications is that they can be customized.

It is also important to note that a company can use the principles of Lean Six Sigma without building an internal training program that offers internal certification. Sometimes, learning and applying DMAIC to projects is enough to show a return on investment in process improvement efforts. Another thing to consider when building an internal program is that most employees can be trained at LSS White for the Yellow Belt level. Whereas a Lean Six Sigma Green belt is valuable to anyone trying to make their job better, faster, or more cost-effective, a company may want to limit that training to employees who will actively participate in the process improvement effort from a cost-effectiveness point of view.

However, if the Lean Six Sigma Black or Master Black belt sees an advantage to building an internal training program, here is a quick reminder of the belts:

- **White Belt**: Understands terminology and program goals.
- **Yellow Belt**: Understands DMAIC and tools.

- **Green Belt**: Can apply to their job or area of expertise.
- Tactical Responsibilities
- Applies Lean Six Sigma to Specific Job
- **Black Belt**: Can apply in all departments.

Strategic Responsibilities

Can Make Process Improvement in any Department
Aligning Projects to Company Goals and Initiatives Risk Analysis
Closed Loop Assessment
Master Black Belt – Mentor, Coach, and Teacher/Facilitator

Remember, Lean Six Sigma does not support radical change management. Science prefers that the employee work in tandem with the existing system. The argument that best supports Lean Six Sigma implementation is that it is scalable, recognizes the importance of speed, and is a quality program that is also cost-effective. It does not believe that everything needs to start from the ground up or that everything needs to be changed.

Tools to express this message include:

- Voice of the Customer (VOC) Emphasis
- Data Collection
- Analysis of Data
- Stakeholders
- Translating into Customer Requirements

To successfully implement Lean Six Sigma, it is often necessary to help people understand the concept of waste. Mistakes are usually agreed upon, but since waste or non-value is often subjective, breaking out the waste areas for discussion is helpful. These include:

- Transportation and Motion
- Over Processing/Over Production
- Defects
- Skills
- Wait Time

Using examples from the workplace when explaining these terms can be amazingly effective. For example, discussing the duplication of specific reports or duties may help the concepts resonate with the employee.

Team building and transparency are essential when applying Lean Six Sigma, which has adopted the popular paradigm that team development follows these phases.

- Forming
- Storming
- Norming
- Performing

This theory believes that successful activity in each phase is necessary to reach the next phase.

This theory is also popular in other business disciplines.

Much attention is paid to collaboration models to implement Lean Six Sigma successfully. These models include:

Win/Win
Collaboration
Agreement
Empathy

These strategies are also helpful in negotiation and conflict management. The learner additionally recognizes the following roadblocks to improvement:

- Lack of Critical Thinking
- Lack of Creative Thinking
- Lack of Basic Project Management Skills
- Poor Communication
- Lack of Motivation

Recognizing these conditions will give the *Leaner* Practitioner a definite advantage when dealing with change management issues.

During the implementation process, it is essential to recognize the various voices mentioned throughout the body of knowledge include:

Voice of the Customer
Voice of the Employee
Voice of the Business (or Industry)
Voice of the Process Voice

The same tools and thinking used in DMAIC can also be used when designing an internal Lean Six Sigma program.

To recap, the following items should be considered for successfully implementing an internal Lean Six Sigma program in a department or enterprise. Lean Six Sigma, sometimes called Lean and Six Sigma, has done a tremendous job increasing Return on Investment. But almost two decades later, it is time for the next generation.

The next generation is SSD Global Solutions product, *Leaner* Six Sigma (L*r*SS©). L*r*SS©, or if you prefer, *Leaner*, simplifies the thoughts, concepts, and terms used in ordinary Lean Six Sigma programs. Using *Leaner*, students can apply models and theories on the fly without constantly consulting huge or complicated manuals. This is because the information is simplified in a way that makes sense the first time it is heard. Then, immediately, the notion is tied to an activity that is meaningful and easy to recall.

This abridged and accelerated approach allows the reader or student more time to work with the context of the issue instead of memorizing convoluted and sometimes outdated tools. With the new push in our society that less education may represent more time to reason things out by discussing issues, *Leaner* is a timely addition to the problem-solving and innovation arena.

Both Six Sigma and Lean Six Sigma are problem-solving models that allow for maximum return on investment by making existing activities better, faster, or more cost-effective by using methodologies and tools designed for those specific purposes.

It is important to note that many organizations informally embrace the basic concepts. Many Lean and Six Sigma tools are now integrated into day-to-day business activities. Used together, Lean and Six Sigma increase the likelihood of success. Since Lean works to eliminate or reduce waste and Six Sigma works to eliminate or minimize mistakes, the methodologies are not mutually exclusive and work well together. This combined use in the industry is called Lean Six Sigma (LSS).

The premise is that LSS, although incredibly successful in its time, is ready for an upgrade. SSD Global Solutions *Leaner* Six Sigma (L*r*SS©) is the next generation.

- Identify existing processes that could be improved, faster, or more cost-effective. This may require assembling a steering committee with cross-functional members.
- Employee training in essential project management

- Employee training in the DMAIC model
- A collaborative understanding of timelines and budgets

Although this chapter introduces two ways to implement *Leaner* Thinking, *Leaner* Manufacturing™ and *Leaner* Six Sigma®, note that any individual can always start working *Leaner* by applying the tools and methodology of the DMAIC. The program can be absent to achieve excellent results.

Appendix 1: Four Quick and Easy Leadership Tips for a Successful *Leaner* Manufacturing™ Business

Tip One – Safety

As the manager of a manufacturing company, it is essential to ensure the safety of your employees. This is achieved by ensuring that all personnel are informed of any modifications to company protocols, upholding safety regulations, and supplying the required personal protective equipment. Employees are likely to experience increased comfort in their work environment when they are assured of their safety. Maintain engagement: To lead a team effectively, it is essential to remain actively engaged with them. Provide input on manufacturing projects and decisions proactively to minimize the need for employees to seek clarification later. The more time you invest now in ensuring that everything is executed correctly. They are minimizing the time required to rectify errors in the future.

Tip Two – Sponsor an Event

Supporting an event is an effective method to connect with your local community. There are numerous opportunities to engage with your local community, ranging from food to 5K races.

Tip Three – Training

Training your employees is essential to maintain your business's success. Here are seven quick tips to help you keep your employees engaged, motivated, and on the same page with company goals

Tip Four – Activities

Here are some low- to no-cost activities:

- Hold regular check-ins.
- Discuss feedback.
- Take time off together.
- Offer flexible work hours.
- Celebrate successes.
- Be transparent about the vision.
- Check in on work/life balance.

Appendix 2: Kaizen/ Rapid Improvement Events in Manufacturing a Step-by-Step Guide

Rapid improvement events (or RIEs) facilitate collaborative efforts among teams to enhance their processes and outcomes, typically within a single day or over a brief period. These events can be highly effective when appropriately executed; however, they are not always simple to coordinate, and numerous businesses encounter challenges in achieving success. This guide will assist you in planning, implementing, and enhancing your initial RIE, ensuring that it optimizes your operations and positions you for greater profitability and growth.

Identify the Issue

Many manufacturing companies strive to lower expenses while enhancing quality and safety standards. Rapid Improvement Events, or RIEs, are structured to assist plant managers in executing enhancements swiftly and with minimal disruption to their operations.

All manufacturing environments can gain advantages from RIEs. The industry you operate is secondary to the extent of change you aim to achieve within a specified time. There are four primary categories of goals:

Choose the Team

Choosing your team is a critical step when initiating a rapid improvement event. You can assemble your team by selecting individuals who are well-versed in the process or by opting for those who have a new perspective. It is crucial to ensure that each team member possesses relevant expertise that they can get to the event. If you seek an individual with a comprehensive understanding of the manufacturing process, it would be advisable to consult a professional from the engineering department. If you seek insights into customer interactions with your product, consider consulting a representative from customer service. Ensure that every team member is dedicated to the manufacturing project and is prepared to engage fully throughout the event.

Establish the Objective

A rapid improvement event, often called a root cause analysis, is a structured brainstorming process that begins with identifying the problem and subsequently focuses on generating a wide array of ideas to address the issue effectively. In manufacturing, these occurrences are utilized when there is a need for prompt problem resolution. RIEs can generate savings of up to 10% in labor expenses. This document provides a comprehensive guide for organizing your own RIE event. Identifying the goal of your event is the initial step you should take. For instance, it would be beneficial to generate strategies to minimize waste from each production line to decrease scrap rates. It is essential to allocate sufficient time for this process a minimum of 2 hours for an average-sized plant or 3–4 hours for a larger facility.

Formulate the Strategy

Identifying the critical areas for enhancement is the initial step. Five essential areas are typically identified: safety, quality, productivity, cost, and customer satisfaction. Upon identifying the critical areas for enhancement, it is necessary to assess your company's current standing in each of these domains and determine the steps needed for improvement. This analysis may be conducted via benchmarking or surveys targeting employees and customers. Upon completing a study, you will identify the areas that require the most focus from your team and determine the starting points for enhancing your business. The subsequent step involves formulating a strategy to tackle these enhancements. A viable approach consists of developing a comprehensive plan that outlines specific goals, methodologies, and metrics to track progress and milestones, enabling the measurement of success effectively. It also outlines the individuals responsible for tasks and their respective deadlines.

Execute the Strategy

Define the objectives and parameters of the event. What motivates you to achieve success? What aspects are you aiming to enhance in the process? What indicators will signify your success? Who are the participants in this event? What requirements do we have for this event? What is the scheduled date for the event? Select a date, time, and location for the enhancement event.

Organize and extend invitations to your participants. Who will be impacted by the results of this event? Encourage individuals to engage in the process to ensure they are informed about the developments and understand that their contributions will have a significant impact.

Ensure that your team is informed about the event s details, including its purpose, start time, and location, so they know whom to expect. Generate a wide range of ideas. It is not necessary to establish a definitive plan or specific concepts.

Begin documenting any ideas or concepts you have regarding the topic. Brainstorming typically involves small groups collaborating without interruption. In the case of an extensive group brainstorming session, one individual should facilitate the discussion while another person documents all suggestions on paper. The leader may inquire about your preferred topic for initial research selection. Which idea do you find yourself feeling the least confident about? Which concept have you encountered the least frequently? Which concept is more likely to achieve success?

Assess the Outcomes

Upon concluding a rapid improvement event, please follow these steps to assess your outcomes.

- Assess whether you have accomplished your objective.
- Assess opportunities for process enhancement and compile a list of actionable items for improvement.

- Execute modifications consistent with the goals and objectives outlined in the organization's strategic plan.
- Inquire with employees about the insights they gained from this experience.
- Conduct a survey to assess stakeholder satisfaction regarding the event.
- Facilitate a post-event debriefing session to address outstanding matters or unresolved concerns.
- Implement an immediate recognition award for team members who have surpassed expectations.
- Record successes and failures to ensure all benefit from these insights into future events.
- Maintain the momentum by organizing another rapid improvement event shortly after the conclusion of the initial one.
- Recognize and acknowledge achievements.

Manufacturing Kaizens, often called Rapid Improvement Events, have gained significant popularity. The concept is straightforward to grasp. Due to several factors, a solution to the problem must be found promptly.

To achieve this, it is often essential to convene all individuals involved in the processes and to possess the authority to make decisions.

On the manufacturing floor, Manufacturing Kaizen frequently occurs in real time. In the context of service or any transactional activity, the objective of Manufacturing Kaizen is typically to efficiently develop a project plan with the budget and timeline already approved. The individuals engaged in the project are consensus regarding the direction of the process improvement.

A variety of models are available for use. A facilitator may possess knowledge of the industry or the specific process improvement or lack familiarity with both. The facilitator guides all meeting participants through a systematic problem-solving plan. A PDCA approach may be practical for straightforward issues. If the problem is more complex, the facilitator will select a DMAIC model. Other Kaizen Manufacturing models have diminished in popularity compared to the DMAIC approach.

Irrespective of the model employed, the objective is to progress swiftly through each phase of the model with the assistance of the facilitator. The facilitator, and at times the project manager, is responsible for

The project generates a project plan informed by participant feedback and secures approval for that plan prior to the team's disbandment.

Manufacturing Kaizens should be regarded as brief improvement initiatives. Effective Kaizen requires thorough preparation and planning in advance. Administrative tasks, including arranging schedules and meeting locations, are essential for success, particularly as participants frequently originate from various regions. Additionally, it is recommended to prepare any pre-reading materials and documentation ahead of the meeting. Manufacturing Kaizens typically lasts between 3 and 5 days.

Traditionally, Manufacturing Kaizens were executed at the C-suite level; however, it is now increasingly common for these initiatives to occur at the mid-management or supervisory level.

The Manufacturing Kaizen is a quality circle comprised of cross-functional members who are invested in the process improvement under consideration. Kaizen is frequently described as having five fundamental elements:

1. Collaboration
2. Self-discipline
3. Enhanced morale
4. Quality circles
5. Recommendations for enhancement

However, given the urgency of the issue being addressed in the Kaizen meeting, these sessions can be intense and demanding. Participants may be required to allocate departmental resources and commit budgeted funds, potentially causing project delays.

Kaizen is a term that combines two Japanese concepts: Kai, meaning change, and Zen, meaning improvement. The term originates from Gemba Kaizen, which emphasizes continuous improvement. The original principle focused on small-scale improvements; however, in recent times, larger issues have been addressed. The cost associated with convening all key human resources to address ancillary issues related to improvement can be significant.

Appendix 3: *Leaner* Six Sigma® in Manufacturing

Many manufacturers have realized the advantage of adding Six Sigma to Lean and practice Lean and Six Sigma. This book referenced Lean Six Sigma, but Appendix 3 is dedicated to how Six Sigma can enhance Lean Manufacturing. Then, this article takes a second step because the author promotes *Leaner* Six Sigma®, not only in manufacturing but in other areas as well. Lean Six Sigma is designed to make it more straightforward to understand and faster for certification, along with a model that allows immediate implementation.

As mentioned in Chapter 10 on implementation, Lean is a methodology that focuses on reducing waste in processes. This is done by looking for ways to make the process more efficient, eliminating unnecessary steps, or simplifying procedures. The goal is to cut costs and deliver better customer service.

No Lean Six Sigma material is complete without explaining some history. The most popular Six Sigma Story is when GE reported $2 billion in savings in 1999 – before that, Toyota claimed the same type of victory using Lean. Both methodologies borrowed heavily from Total Quality Management (TQM). Initially, Six Sigma concentrated on eliminating mistakes and defects, and Lean focused on removing waste. Eventually, most employers agreed that Six Sigma and Lean could be used together. Since both had complementary tools and the same intent to make things better, faster, and more cost-effective, the two programs were often combined.

Six Sigma is a statistical measurement based on Defects per Million Opportunities (DO). A defect is defined as any nonconformance of quality. At Six Sigma, only 3.4 DMPO may occur. To use sigma as a measurement, there must be something to count, and everyone must agree on what constitutes a defect. Regular distribution models look at three Sigma, which is essentially 6,210 DO. Some processes are acceptable at lower sigma levels; in many cases, 6 Sigma is considered ideal. Sigma (σ) is a symbol from the Greek alphabet used in statistics when measuring variability. The Six Sigma methodology measures a company's performance by the sigma level. Sigma levels are a measurement of error rates. It costs money to fix errors, so saving on this expense can be directly transferred to the bottom line.

Typical tools taught in a Six Sigma include:

- Process Mapping
- Affinity Diagram or KJ (Kawakita Jiro) method
- Measurement System Analysis or MSA
- Pareto Chart
- SIPOC Analysis – Supplier-Input-Process-Output-Customer
- Scatter Diagram or Scatter Plot
- Quality Function Deployment or QFD, also known as House of Quality
- Ishikawa Diagram or Fishbone
- Failure Mode Effects Analysis or "FMEA"
- Failure Mode and Effects Criticality Analysis or "FMECA"
- Value Chain Map
- Histogram
- Control Plan

Lean Manufacturing is a production practice that concentrates on eliminating waste. It is based on the total production system (TPS), introduced originally by Toyota, and the principles of TQM. TQM capitalizes on the involvement of management, workforce, suppliers, and customers to meet or exceed customer expectations.

Initially, Lean identified the following as the worst forms of waste:

- Transportation
- Inventory (all components, work in progress, and finished product not being processed)
- Movement
- Waiting
- Over Production
- Over Processing
- Defects
- Skills

An easy way to remember the primary waste is T-I-M W-O-O-D-S.

Eventually, Lean evolved to consider additional forms of waste. Lean Thinking is designed to:

- Shrink lead times
- Save turnover expenses
- Reduce setup times
- Avoid unnecessary expenses
- Increase profits

Lean focuses on getting the right things to the right place at the right time and quantity while minimizing waste. Lean makes the work simple enough to understand, do, and manage. The very nature of Lean would suggest that it would be "wasteful" to spend time trying to understand manuals or complicated processes, so it is best to simplify the language.

Typical tools promoted in Lean include:

- 5S
- Error Proofing
- Current Reality Trees
- Conflict Resolution Diagram
- Future Reality Diagram
- Inventory Turnover Rate
- JIT – Just-in-Time Theories
- Kaizen
- Kanban
- Lean Metric
- One-piece Flow
- Overall Effectiveness
- Prerequisite Tree
- Process Route Table
- Quick Changeover
- Standard Rate or Work
- Takt Time
- Theory of Constraints
- Total Productive Maintenance
- Toyota Production System
- Transition Tree
- Value-added to non-value-added lead time ratio

- Value Stream Mapping
- Value Stream Costing
- Visual Management
- Workflow Diagram

Another variation of Lean is Lean Office. There are seven primary principles of a Lean Office. These are continually being updated

1. Committed Leadership
2. Establishing Metrics and Goals
3. Standardized Processes
4. 5S – a physical organizational system
5. Minimal Work in Progress (WIP)
6. Positive Workflow
7. An Understanding of Demand

Lean offices use a daily management system and communication like Agile. They are typically visual, with signs pointing the way and labels on cabinets. Finally, they foster a culture of continuous improvement.

Lean Six Sigma began in the late 1990s. Six Sigma and Lean have already expanded to include service and manufacturing. Allied Signal and Maytag independently started experimenting with using both methodologies. Employees were cross-trained. Maytag was the first to recognize that Lean manufacturing and Six Sigma do not conflict with each other and are perfect complements. The United States Army shares this thinking. Everyone involved in the movement to implement Lean Six Sigma saw the power of combining both toolboxes and attacking both defects and wastes. Many shared tools are apparent in both methodologies that reduce the learning curve.

Lean Six Sigma can be used in any industry, including finance, Construction, Government, Healthcare, Insurance, and Hospitality. It is about increasing quality and profit. The new tools include methodologies based on teamwork as a principle. Process improvement is not linear; each component is handed off to another department or individual. Each member of the process is involved in improving client satisfaction.

The new *Leaner* tools focus on Continuous Improvement as a guiding principle. The road to quality is paved with minor incremental improvements. Significant sweeping changes seldom work. As this country moves its business style from control to management to leadership, we find that the people doing the work can most identify changes necessary to improve quality. Leadership must listen and implement changes rather than direct the solutions. Some examples are:

- Improving Forecast Accuracy
- Reducing Volume of Rejected Orders
- Improving Consumer Loan Cycle Time
- Reducing Engine Installation Times
- Eliminating Mistakes in an Operating Room
- Reducing Pharmacy Dispensing Error Rates
- Improving the Effectiveness of Employee Hazard Recognition
- Reducing Process Variation Costs Related to Manufacturing

Before examining Lean Six Sigma, Continuous Improvement (CI) programs should be explored. Most companies have some quality control programs. These programs may be formal or informal. Some programs have defined documents and manuals; other quality programs are not recorded or tracked. Quality is a large piece of continual improvement. All CI programs ask two questions: (1) who are the customers? and (2) what will it take to satisfy them?

Both Lean and Six Sigma endorse the PDCA model. PDCA stands for Plan, Do, Check, Act. This popular project management tool is easy to understand. It is also called the Deming Wheel or Deming Cycle.

- **Plan**: Identify an Opportunity and Plan for Change.
- **Do**: Implement the Change on a Small Scale.
- **Check**: Use Data to Analyze the Results of the Change and Determine Whether It Made a Difference.
- **Act**: If the Change Was Successful, Implement It on a Wider Scale and Continuously Assess the Results. If the Change does not Work, Begin the Cycle Again.

Lean and Six Sigma support continuous improvement (CI), an ongoing effort to improve products, services, or processes. CI can be incremental (over time) or breakthrough (all at once). CI programs are often not proactive and present with a problem up front. Within any problem-solving model, there are four steps to remember: Define the Problem, Generate the Solution, Evaluate and Select an Alternative, and Implement.

Lean Six Sigma uses a set of quality tools that are often used in TQM. These tools, sometimes referred to as problem-solving tools, include:

- Control charts
- Pareto Diagrams
- Process Mapping
- Root Cause Analysis
- SPC

As noted throughout this book, Lean Six Sigma largely depends on the DMAIC (Define-Measure-Analyze-Improve-Control) model (Define, Measure, Analyze, Improve, and Control) developed by Motorola and later enhanced by General Electric. Some historical documents claim that GE was responsible for the creation or at least the modification of the Define Phase.

- Define the process improvement goals consistent with customer demands and enterprise strategy.
- Measure the current process and collect relevant data for future comparison.
- Analyze the relationship between factors. Determine the relationship and attempt to ensure that all aspects have been considered.
- Improve or optimize the process based on the analysis.
- Control to ensure that any variances are corrected before defects occur.

Lean Six Sigma uses the tools above. It also uses the martial arts designations White, Yellow, Green, Black, and Master Black Belt to denote the level of expertise. The generally accepted belt designations are as follows:

- **White Belts**: individuals who have been given a basic orientation.
- **Yellow Belts**: individuals trained in applying Six Sigma management tools.
- **Green Belts**: individuals who handle Lean Six Sigma implementation and other regular job responsibilities.
- **Black Belts**: individuals who devote 100% of their time to Lean Six Sigma initiatives.
- **Master Black Belts**: individuals who act in a teaching, mentoring, and coaching role.

The roles, responsibilities, and education for Lean Six Sigma Green Belts and Black Belts are the most consistent. The significant difference between a Green Belt and a Black Belt is that Green Belts have a regular job where they apply process improvement via Lean Six Sigma. In contrast, Black Belts are engaged solely in process improvement efforts. Individual expertise must be blended with Lean Six Sigma tools and methods to be a successful Green Belt.

To be a successful Black Belt or Master Black Belt, one must have an overall understanding of business, usually in combination with a Master of Business Administration degree. Information Technologists, who by nature must interface with various departments or individuals who have owned their businesses, are often excellent candidates for Lean Six Sigma Black Belt studies. Black Belts and Green Belts need a valid understanding of essential project management.

Other roles in the Lean Six Sigma organization include Sponsor, Process Owner, and Cross-Functional Teams. The sponsor is generally the person paying for the project. The process owner usually is responsible for the process's success. The cross-functional team is the ideal team promoted by Lean Six Sigma – a team of multiple disciplines, including functional expertise, finance, marketing, and operations.

The roles and responsibilities in Lean Six Sigma are still rooted in TQM. In a TQM effort, all members of an organization participate in improving processes, products, and services. TQM practices are based on cross-functional product design and process management. Other components related to Lean Six Sigma also covered in TQM include:

- Supplier relations
- Overall quality management
- Customer and employee involvement
- Information
- Feedback
- Committed leadership
- Strategic planning

The *Leaner* Practitioner interested in better understanding the history of process improvement will benefit from studying material first presented within the TQM framework. Works by

Edwards Deming and Joseph Juran are still prevalent today. Other significant authors include Kaoru Ishikawa, A.V. Feigenbaum, and Phillip B. Crosby.

We aim to simplify everything to provide a brief history of *Leaner* Six Sigma®. This includes the model, the tools, the process, and the presentation of results. We believe that much of the work in Lean Six Sigma has already been accomplished. Therefore, it might be okay, especially in certain circumstances, to use an MS Excel calculator, a template, automated tools, or someone else's expertise when completing a project. *Leaner* believes in the collective intelligence of teams, sharing information, and learning from each other. In other words, sometimes you feel like making a cake from scratch, but sometimes, using a cake mix or simply purchasing the pastry makes more sense. How the results are achieved is not nearly as important as achieving the results.

To a certain extent, *Leaner* believes that it is not always necessary to work hard to get meaningful results – sometimes it's just about working smarter, paying attention to what is already available in tapping into your critical thinking skills.

Applying essential process improvement in the workplace appears primarily in the following examples:

- Mapping the processes
- Analyzing the processes
- Redesigning the processes
- Acquiring the right resources
- Implementing and communicating the change
- Reviewing the processes for results

Leaner Manufacturing™ Key Terms – Simplified

These terms are designed as high-level explanations of various Lean and Manufacturing terms and are intended for quick reference that is simplified.

M's – Four-word categories are used to provoke thought in an Ishikawa diagram (cause & effect diagram): material, method, machine, and man.

Whys – The practice of asking why five times when presented with a problem to identify potential root causes.

5S – A process and method for creating and maintaining an organized, clean, safe, and high-performing workplace. The steps are Sort, Set, Shine, and Standardize and Sustain.

A3 Report – A Toyota-developed standard report shows a problem, analysis, and corrective action plan on a single piece of paper, usually A3 size.

A-B Control – A method used to regulate working relationships between a pair of operations such that overproduction is minimized. Machine A cannot feed machine B until it is empty or waiting for work.

ABC Production Analysis – Volume-based groupings segment part numbers and the inventory policy within which they fall.

Abscissa – The horizontal axis of a graph.

Acceptance Criteria – The customer identifies specific criteria for each functional requirement. The acceptance criteria are written in simple terms from the customer's perspective.

Acceptance Testing – Acceptance testing is a validation activity determining whether a system satisfies its acceptance criteria. It is the last phase of the software testing process.

Acceptance Region – The region of values for which the null hypothesis is accepted.

Affinity Diagram – This organizes brainstorming ideas into categories or themes. It is useful when large amounts of information are collected during brainstorming. It is also called the KJ method after Kawakita Jiro (a Japanese anthropologist) first developed the idea.

Agile – A conceptual framework for undertaking software manufacturing projects. Agile methods are not a single approach.

Alpha Risk – The probability of accepting the alternate hypothesis when the null hypothesis is true.

Alternative Hypothesis – A tentative explanation indicating that an event does not follow a chance distribution, contrasting the null hypothesis.

Andon – is the Japanese word for signal, referring to a visual system that indicates supervision when abnormalities occur within processes.

ANOVA – Analysis of variance. This statistical test is done by comparing the variances around the means of the condition being compared. In its simplest form, ANOVA gives a statistical test of whether the means of several groups are all equal.

Anova Gauge R&R – Measures the amount of variability induced in measure by the measurement system itself and compares it to the total variability observed to determine the viability of the measurement system.

Assemble to Order – An environment where a product or service can be assembled or provided upon receipt of a customer's order. The product will usually consist of several modules assembled to the highest level possible and stored so that when the order arrives, it can be assembled quickly and to the customer's specifications.

Assembly Line – Most industrial processes include some assembly lines. According to Wikipedia, "An assembly line is a manufacturing process in which parts (usually interchangeable parts) are added as the semi-finished assembly moves from workstation to workstation where the parts are added in sequence until the final assembly is produced. By mechanically moving the parts to the assembly work and moving the semi-finished assembly from workstation to workstation, a finished product can be assembled faster and with less labor than by having workers carry parts to a stationary piece for assembly."

Assignable Cause – A source of variation that is nonrandom; a change in the source ("Vital Few" variables) will produce a significant change of some magnitude in the response (dependent variable), e.g., a correlation exists; the change may be due to an intermittent in – phase effect or a constant cause system which may or may not be highly predictable; an assignable cause is often signaled by an excessive number of data points outside limits; an unnatural source of variation; most often economical to eliminate.

Assignable Variation – Variation in data which can be attributed to specific causes.

Assumption – There may be external circumstances or events that must occur for the manufacturing project to be successful (or that should happen to increase your chances of success). If you believe that the probability of the event occurring is acceptable, you could list it as an assumption. An assumption has a probability between 0% and 100%.

Attribute – A characteristic that may take on only one value.

Attribute Data – Numerical information at the nominal level; subdivision is not conceptually meaningful; data representing the frequency of occurrence within some discrete category, e.g., 42 solder shorts.

Automatic Line Stop – Ensuring production processes will stop whenever a defect or problem occurs.

Autonomation – A term developed by Taiichi Ohno to describe "automation with a human touch." Automated machines will stop when abnormalities occur so they will not create large amounts of scrap and do not need an operator to watch the machine.

Availability % – is when a machine is available to produce a product and is given as a percentage of the total planned production time. A manufacturing analytics platform can determine the availability of a machine by collecting and calculating downtime using alarms and signals from controls on the equipment and operator input.

Average – Also called the mean, it is the arithmetic average of all sample values. It is calculated by adding all the sample values and dividing by the number of elements (n) in the sample.

Background Variables – Variables are of no experimental interest and are not held constant. Their effects are often assumed insignificant or negligible, or they are randomized to ensure that contamination of the primary response does not occur.

Balance Chart – A bar chart or histogram illustrating work content per operator. It can balance work for operators or machines to improve flow.

Balanced Scorecard – is a performance management approach that focuses on the customer perspective, internal business processes, learning and growth, and financials.

Batch and Queue – This is a typical mass production method in which a part going through a system is produced in large batches to maximize "efficiency" and then sits in a queue waiting for the next operation.

Benchmarking – A standard used to compare performance against best-in-class companies. It then uses the information gathered to improve its performance. Subjects that can be benchmarked include strategies, products, programs, services, operations, processes, and procedures.

Beta Risk – The probability of accepting the null hypothesis when the alternate hypothesis is true.

Bill of Materials – In manufacturing, a bill of materials (BOM) is all of the materials and parts needed to complete a particular product.

Blocking Variables – A relatively homogenous set of conditions within which different conditions of the primary variables are compared. Used to ensure that background variables do not contaminate the evaluation of the primary variable.

Bottleneck – A bottleneck is a sort of congestion in a system that occurs when workload arrives at a given point more quickly than that point can handle it. It is metaphorically derived from the flowing of water through a narrow-mouthed bottle where the flow of water is constrained by the size of its neck.

Breakthrough Improvement – The rate of improvement at or near 70% over baseline performance of the as-is process characteristics.

Brownfield – A brownfield site is an existing facility usually managed in line with mass production methods.

Bugs – A software bug is a problem that causes a program to crash or produce invalid output. It is caused by insufficient or erroneous logic and can be an error, mistake, defect, or fault.

Build to Order – A production environment where a product or service can be made and assembled after receipt of a customer's order.

Burndown Chart – A burndown chart is a visual tool for measuring and displaying progress. Visually, a burndown chart is simply a line chart representing remaining work overtime. Burndown charts measure the progress of an agile manufacturing project at both the iteration and manufacturing project levels.

Capability – Measurement index that expresses the capability of the process by using a percentage. Or, even simpler, the ability to do something.

Capital Linearity – A philosophy linked to capital expenditure on machinery such that a small amount of additional capacity can be added using several smaller machines rather than one big and costly machine.

Casualty – The principle that every change implies the operation of a cause.

Causative – Effective as a cause.

Cause – That which produces an effect or brings about change.

Cause and Effect Diagram – Also called a fishbone diagram. This graph places the issue being discussed in the head of the fish. The bones of the fish are categories of problems that could be a problem. The smaller bones are the possible root causes.

Cell – A cell is a group of people, machines, materials, and methods arranged so that processing steps are located adjacent and in sequential order. This allows parts to be processed one at a time or, in some cases, in a constant small batch maintained through the process sequence. The purpose of a cell is to achieve and maintain an efficient, continuous flow of work.

Center Line – The line on a statistical process control chart which represents the characteristic's central tendency.

Central Tendency – data clustered around the middle. Mean, mode, and media are all examples.

Cell – The term cell is used in the larger concept of cellular manufacturing which is when equipment is arranged in small increments, or a cell, to promote continuous flow production. Most often in a U-shape, the cell is designed to facilitate shorter production times and increased efficiencies between each step.

Chaku-Chaku – One-piece flow ideal whereby machines automatically unload parts so that an operator can move apart from one machine to the next without stopping to unload parts.

Champion – A person who supports the successful completion of the manufacturing project.

Characteristic – A process input or output which can be measured and monitored.

Classification – Differentiation of variables.

Client/Customers – The person or group directly beneficiary of a manufacturing project or service is the client/customer. These are the people for whom the manufacturing project is undertaken (indirect beneficiaries are stakeholders). In many organizations, internal beneficiaries are called "clients," and external beneficiaries are called "customers," but this is not a hard and fast rule.

CMM – Coordinate-Measuring Machine ensures measurements of a product are correct. If you're making something out of metal and it has a really tight tolerance, for example if it needs to be perfectly flat, the CMM checks to make sure the measurements are in line with the standard.

Common Cause – SEE Random Cause.

Common Causes of Variation – Sources of variability in a genuinely random process. These are generally inherent in the process itself and can be managed. This type of variation is a usual, historical, quantifiable variation in a system.

Complexity – The difficulty level to build, solve, or understand something based on the number of inputs, interactions, and uncertainty involved.

Confidence Level – The probability that a random variable x lies within a defined interval.

Confidence Limits – These are the two values that define the confidence level.

Confounding – Allowing two or more variables to vary together so that it is impossible to separate their unique effects.

Constraints – Constraints are limitations that are outside the control of the manufacturing project team and need to be managed around.

Consumer Risk – Probability of accepting a lot when, in fact, the lot should have been rejected (see Beta Risk).

Continuous Data – A set of observations usually associated with a physical measurement that can take on any mathematical value within specified parameters. More simply put, it is the sequencing of activities.

Continuous Flow – Each process, whether in an office or plant setting, makes or completes only the one piece the following process needs; the batch size is one. Single-piece or one-piece flow is the opposite of a batch-and-queue process.

Continuous Random Variable – A random variable that can continuously assume any value in some specified variable.

Control Chart – The most powerful tool of statistical process control. It comprises a run chart, statistically determined upper and lower control limits, and a centerline.

Control Limits – Upper and lower bounds in a control chart are determined by the process. They can be used to detect special or common causes of variation.

Control Specifications – Specifications called for by the product being manufactured.

Cost of Poor Quality (COPQ) – The costs associated with any activity that does not do the right thing the first time.

Critical Path – The series of consecutive activities representing the process's most prolonged time path. The critical path is the sequence of activities that must be completed on schedule for the entire manufacturing project to be completed on schedule. It is the longest-duration path through the work plan. Suppose an activity on the critical path is delayed by one day. In that case, the entire manufacturing project will be delayed by one day (unless another activity on the critical path can be accelerated by one day).

Critical to Quality (CTQ) – Any activity or thought related to the successful outcomes of the manufacturing project.

Cross Dock – A facility that gathers and recombines a variety of inbound materials and parts from multiple suppliers to forward to various customers.

Cutoff Point – The point that partitions the acceptance region from the rejection.

Cycle Efficiency (CE) – CE is a measure of the relative efficiency in a production system. It represents the percentage of value-added time of a product through the critical path vs. the total cycle time (TCT).

Cycle Efficiency (CE) – CE measures the relative efficiency of a production system. It represents a product's value-added time percentage through the critical path vs. the total cycle time (TCT).

Cycle Time Interval – The frequency at which an item is made during a set period (usually days).

Cycle time – is when a person needs to complete an assigned task or activity before starting again.

Daily Standup/Scrum – A Daily Standup is a whole team meeting that happens at the same time every day and usually lasts 15 minutes or less. Each team member should provide the following information: what did I do yesterday, what am I planning to do today, and what impediments do I currently have?

Data – Information used as a basis for reasoning, discussion, or calculation; often refers to quantitative information.

Defect – An output of a process that does not meet a defined specification, requirement, or desire such as time, length, color, finish, quantity, and temperature.

Defective – A product or service unit containing at least one defect.

Degrees of Freedom – The number of independent measurements for estimating a population parameter.

Deliverable – A deliverable is any tangible outcome that is produced by the manufacturing project. All manufacturing projects create deliverables. These can be documents, plans, computer systems, buildings, aircraft, etc. Internal deliverables are produced by executing the manufacturing project and are usually needed only by the manufacturing project team. External deliverables are those that are created for clients and stakeholders. Your manufacturing project may create one or many deliverables.

Demand – The usage of an item over a period. This also includes understanding the customer requirements for quality, lead time, and price.

Density Function – The function that yields the probability that a random variable takes on any one of its possible values.

Dependent Variable – A response variable, e.g., y is the dependent or "Response" variable where Y = f (XI…Xn) variable.

Design of Experiments (DOE) – An efficient, structured, and proven approach to interrogating a process or system to maximize the gain in process or system knowledge.

Design for Six Sigma (DFSS) – The use of Six Sigma thinking, tools, and methods applied to the design of products. Any Six Sigma model for managing a manufacturing project that is not DMAIC is generally considered a DFSS.

Discrete Random Variable – A random variable that can assume values only from a definite number of discrete.

Distributions – Tendency of large numbers of observations to group themselves around some central value with a certain amount of variation or "scatter" on either side.

DMAIC – This acronym stands for "Define, measure, analyze, improve, and control." It is the heart of the Six Sigma process and refers to a data-driven quality strategy for improving processes. It is an integral part of any company's Six Sigma quality initiatives.

Downtime – the time a plant or particular cell, line, or machine is down or offline and not producing any product.

Effect – That which was produced by a cause.

Engineer to Order – Products whose customer's specifications are unique for each order; therefore, each product is engineered from scratch upon receipt of an order.

Epic – A huge user story that is eventually broken down into smaller stories.

Estimation – The process of agreeing on a size measurement for the stories and the tasks required to implement those stories in a product backlog.

Exits – The work completed over time measured in £'s or Units.

Experiment – A test under defined conditions to determine an unknown effect; to illustrate or verify a known law; test or establish a hypothesis.

Experimental Error – Variation in an observation made under identical test conditions. It is also called residual error. The amount of variation cannot be the variables included in the experiment.

Factors – Independent variables.

Failure Mode and Effects Analysis (FMEA) – A procedure used to identify, assess, and mitigate risks associated with potential products, system, or process failure modes.

Feature Creep – occurs when software becomes complicated and challenging to use due to having too many features.

FIFO – A strategically sized inventory that keeps the sequence of the production uniform throughout the value Stream maintaining flow.

Finish to Order – An environment such that products are built to a higher level as possible, and then configured to customers' requirements, upon receipt of order.

Finished Goods – In a manufacturing environment, finished goods are products that have completed all stages of production and are now ready to be distributed to customers. Calculating and reporting finished goods inventory helps manufacturers prevent waste by producing too much, calculate profitability once finished goods have been sold or distributed, and improve the production process.

Fishbone Diagram – See the cause-and-effect diagram.

Fixed Effects Model – Experimental treatments are specifically selected by the researcher. Conclusion only applies to the factor levels considered in the analysis. Inferences are restricted to experimental levels.

Fixed-Position Stop System – This is a problem-addressing method for continuously moving production lines. If a problem is identified and not resolved before a fixed point, the production line will stop.

Flowchart – A graphic model of the flow of activities, material, and/or information during a process.

Fluctuations – Data variances caused by many-minute variations or differences.

Frequency Distribution – The pattern or shape formed by the group of measurements in a distribution.

Gage R&R – is used in Measurement Systems Analysis (MSA) Quantitative assessment of how much variation (repeatability and reproducibility) is in a measurement system compared to the total variation of the process or system.

Gantt Chart – A Gantt chart is a bar chart that depicts activities as blocks over time. The beginning and end of the block correspond to the start and end date of the activity.

Gemba – A Japanese term used to describe the "actual place" where value is added or the shop floor.

Greenfield – A new production facility not restricted by practices of the past therefore having a culture of adapting change without resistance.

Heijunka – Leveling production by product and/or quantity over a fixed period.

High-Level Value Stream Map – A visual representation of the aggregated material and information flows within a company or business unit.

Histogram – A bar chart that depicts the frequencies (by the height of the plotted bars) of numerical or measurement categories.

HMI – Human-Machine Interface is self-explanatory, but the user interface connects a human user to a machine. For example, the Mingo platform has an HMI component with the operator interface we provide to machine operators to input data during manufacturing.

Homogeneity of Variance – The variances of the contrasted groups are equal (as Defined by statistical test of significant differences).

Hoshin – is the Japanese word for planning used throughout operational, financial, strategic, and manufacturing project-based scenarios.

Independent Variable – A controlled variable whose value is independent of the value of another variable.

Input – A resource consumed, utilized, or added to a process or system. Synonymous with X, characteristic, and input variable.

Inspection – Mass production would use inspectors outside of a process. Lean producers assign quality responsibility to the areas where the processes are performed. Inspections are performed within the areas that own the assembly process.

Instability – Unnaturally large fluctuations in a pattern.

Interaction – The combined effect of two factors observed over and above the singular impact of each factor against the other factor's level. A significant interaction indicates that the impact of each factor on the response changes depending on the value of the other factor.

Interval – Numeric categories with equal units of measure but no absolute zero point, i.e., quality scale or index.

Inventory Turns – A measure to quantify the pace at which inventory rotates throughout a company. Inventory turns = annual cost of goods sold / average inventory value.

Jidoka – Quality is built into processes, so if a process cannot create the required output, it will not operate until it can.

Jishuken – is a Japanese word describing a "hands-on learning workshop."

Just in Time (JIT) – stands for "just in time." This means producing or conveying only the items needed by the following process when needed and in the required quantity. This process can even be used between facilities or companies.

Kaikaku – Radical improvement designed to quickly eliminate and/or add value to a value stream. Also described as Breakthrough Kaizen.

Kaizen – Incremental change for the better. The organized use of common sense to improve cost, quality, delivery, safety, and responsiveness to customer needs.

Kanban – Kanban, pronounced /'kɑn'bɑn/, is a method for developing products with an emphasis on just-in-time delivery and the optimization of the flow of work on the team. It emphasizes that developers pull work from a queue, and the process, from the definition of a task to its delivery to the customer, is displayed for participants to see.

Kanban post – A storage container for Kanban cards pulling deliveries.

Labor Linearity – A manning philosophy such that as demand increases or reduces manpower is added one at a time; as such manpower requirements are linear to production volume.

Lead-Time – The total time from the beginning of the supply chain to the time something needs to ship. The sum of the VA/NVA time for a product to move through the entire value stream.

Lean Transactional – The application of lean to business processes, such as paperwork flow through an office in accounts or marketing.

Level Selling – The elimination of sales spikes generated by end-of-month sales targets at dealers. This allows for improved customer demand flow and improvements in anticipated demand.

Lifecycle – refers to the process used to build the deliverables produced by the manufacturing project. There are many models for manufacturing project lifecycles.

Line – Charts track performance without a relationship to the process capability of control limits.

Linear regression – Analyze the relationship between two variables, X and Y.

Long-term Variation – The observed variation of an input or output characteristic that has had the opportunity to be observed over time.

Lower Control Limit (LCL) – Used in control charts to show the lower limit. Typically, three standard deviations below the central tendency.

Machine Cycle Time – The amount of time a unit spends on the operational cycle of a machine.

Make-to-Assemble (MTA) – There are a couple of different types of manufacturing production processes. A make-to-assemble (MTA) strategy focuses on having readily available parts but does not start making the finished product until an order has been placed. This strategy allows for some customization but not full customization like the make-to-order strategy.

Make-to-order (MTO) – In manufacturing, this is a process where a product is only made after a customer has placed an order. Typically, these are highly customized and personalized products made to fit a customer's specifications. Think about a custom piece of furniture or even a custom car – these are both examples of a made-to-order (MTO) product.

Make to Stock (MTS) – This is a traditional strategy many manufacturers follow. Make-to-stock (MTS) creates inventory in anticipation of customer demand. This strategy requires an accurate forecast, which is not always easy. If the forecast is wrong, manufacturers can be left with too much inventory or too little.

Manufacturing Project – A manufacturing project is a unique and temporary structure that organizes and manages work to build a specific Defined deliverable or set of deliverables. Each project's uniqueness underscores its significance and the difficulty of comparing different manufacturing projects.

Manufacturing Project Phase – A phase is a major logical grouping of work on a manufacturing project. It also represents the completion of a significant delivery or set of related deliverables.

Manufacturing Project Team – The manufacturing project team comprises the full-time and part-time resources assigned to work on the project's deliverables. They are responsible for understanding the work to be completed and completing assigned work within the budget, timeline, and quality expectations.

Mean – The statistical measure of a sample used to estimate the meaning of the population from which the sample was drawn. Numerically, it equals the sum of scores divided by the number of samples.

Measurement Accuracy is used for repeated measurements – It is a comparison of the average of the measurements to some known standard.

Measurement Precision – is the variation in the measured values for a repeated measurement.

Median – The middle value of a data set when the values are arranged in ascending or descending order.

Metric – A measure that is a key indicator of performance. It should be linked to goals or objectives and carefully monitored.

Milestone – A milestone is a scheduling event that signifies the completion of a significant delivery or related deliverables. A milestone, by definition, has a duration of zero and no effort. There is no work associated with a milestone. It is a flag in the work plan to signify that some other work has been completed. Usually, a milestone is used as a manufacturing project checkpoint to validate how the manufacturing project is progressing. In many cases, a decision, such as validating that the manufacturing project is ready to proceed further, must be made at a milestone.

Milk Run – Reducing transport costs and batch sizes by performing multiple pick up and drops at multiple suppliers using one truck.

Mixed Effects Model – Contain elements of both the fixed and random effects models.

MTBF – Mean Time between failures, which is the expected time between inherent failures of a machine or machinery. It is calculated by averaging the time between those failures. However, the definition of failure completely depends on each company, as a failure may be defined differently from one company to the next. If you have recently switched jobs, the methodology for calculating failure will remain the same, but the factors for determining a failure may differ.

MTTR – Mean Time to Repair – MTTR and MTBF go hand-in-hand because while MTBF calculates the average time between failures, MTTR calculates the average time it takes to repair that failure. Again, this calculation varies from one company to the next.

Muda – The Japanese word for waste or non-value-added activity.

Mura – The Japanese word used to describe variation or fluctuation.

Muri – The Japanese word used to describe overburdening or strain/stress.

Nemawashi – A Japanese expression used to describe the practice of obtaining support and buy-in for change by firstly the idea and then the plan with upper management and stakeholders. Directly translated, it means "preparing" the ground for planting.

Nominal – Unordered categories that indicate membership or non-membership with no implication of quantity, i.e., assembly area number one, part numbers, etc.

Nonconforming Unit – A unit that does not conform to one or more specifications, standards, and/or requirements.

Nonconformity – A condition within a unit that does not conform to some specific specification, standard, and/or requirement, often referred to as a defect; any given nonconforming unit can have the potential for more than one nonconformity.

Non-Value Added (NVA) – Any activity performed in producing a product or delivering a service that does not add value.

Normal Distribution – The distribution characterized by the smooth, bell-shaped curve.

Null Hypothesis – A tentative explanation that indicates that a chance distribution is operating, in contrast to the null hypothesis.

Obeya – Translated as "Big Room" this is the expression used by the Japanese to describe the powerful manufacturing project room concept also known as a "war room."

Objective – An objective is a concrete statement that describes what the manufacturing project is trying to achieve. The objective should be written at a low level, so that it can be evaluated at the conclusion of a manufacturing project to see whether it was achieved.

OEE – Overall Equipment Effectiveness, the most well-known metric in the manufacturing industry, evaluates availability, performance, and quality to determine machine effectiveness. Using a formula developed by the automotive manufacturer Toyota, OEE is calculated by multiplying all of those factors together to get one single metric. Learn more about and the production of goods.

One-Piece Flow – Making and moving only one piece or part at a time. See Continuous Flow.

One-sided Alternative – The value of a parameter which has an upper bound and a lower bound, but not both.

OPC – Open Platform Communications which is a series of standards that apply to industrial telecommunication, OPC specifies communication of real-time plant data between control devices from different manufacturers.

Operator Cycle Time – The time it takes an operator to go through all his or her work elements before repeating them.

Order Interval – This represents the frequency (days) that a part is ordered.

Ordinal – Ordered categories (ranking) with no information about the distance between each category, i.e., the rank ordering of several measurements of an output parameter.

Ordinate – The vertical axis of a graph.

OT – Operational Technology – Used by hands on the factory floor, OT is the software that changes machine processes in a plant or factory. For example, the software can control the use of valves or pumps.

Overproduction – The process of producing more, sooner, or faster than the next process or customer requires.

P Charts – Charts used to plot percent defectives in a sample.

Pacemaker – The only point in the production process that is scheduled and, therefore, dictates the pace of production for a whole system of processes.

Pacesetter – The point in the process that limits the output of the total process.

Parameter – A constant defining a particular property of the density function of a variable.

Pareto Diagram – A chart that ranks or places everyday occurrences in order.

Performance % – Often referred to as throughput, performance measures the actual cycle time versus the ideal cycle time. The performance of a machine, cell, or line indicates the ability to meet the schedule or deliver to customers on time. Manufacturing analytics software can calculate performance by collecting part or product counts from the machines and comparing them against the ideal cycle time in the system.

Perturbation – A nonrandom disturbance.

Pitch – The amount of time required by a production area to make one container of products. Takt time × pack-out qty = pitch.

Plan for Every Part (PFEP) – A comprehensive plan for each part consumed within a production process. This would take the form of a spreadsheet or simple table containing data such as pack-out quantity, location of use and storage, order frequency, etc. This provides one accurate source of information relating to parts.

Plan, Do, Check, Act (PDCA) – An improvement cycle introduction to the Japanese in the 50s by W. Edwards Deming. Based upon proposing then implementing an improvement, measuring the results, and acting accordingly.

PLC – stands for Programmable Logic Controller. Basically, it is a computer that has been modified or programmed to be used for the manufacturing process. The controllers can specifically automate production processes, machine functions, or even an entire line.

Poka Yoke – Mistake-proof device or procedure designed to prevent a defect from occurring throughout the system or process.

Population – A group of similar items from which a sample is drawn and often referred to as the universe.

Power of an Experiment – The probability of rejecting the null hypothesis when it is false and accepting the alternative hypothesis when it is true.

Prevention – The practice of eliminating unwanted variations of priori (before the fact), e.g., predicting a future condition from a control chart and when applying corrective action before the expected event transpires.

Primary Control Variables – The major independent variables in the experiment.

Probability – The chance of something happening in percentage or number of occurrences over many trials.

Probability of an Event – The number of successful events divided by the total number of trials.

Problem – A deviation from a specified standard.

Problem-Solving – A process of solving problems; the isolation and control of those conditions which generate or facilitate the creation of undesirable symptoms.

Process – A method of doing something involving several steps or operations.

Process Average – The central tendency of a given process characteristics across time or a specific point.

Process Control Chart – Any of several various types of graphs upon which data are plotted against specific control limits.

Process Owner – I am responsible for process performance and resources. They provide support, resources, and functional expertise to Six Sigma manufacturing projects. They are accountable for implementing developed Six Sigma solutions into their process.

Process Spread – The range of values a given process characteristic displays; this term most often applies to the range but may also encompass the variance. The spread may be based on data collected at a specific time or reflect the variability across a given time.

Producers Risk – Probability of rejecting a lot when, in fact, the lot should have been accepted (see Alpha Risk).

Product Backlog – This acts as a repository for requirements targeted for release at some point. These are typically high-level requirements with high-level estimates provided by the product stakeholders. The requirements are listed on the backlog in priority order and maintained by the product owner.

Product Owner – The product owner, as the voice of the customer, plays a crucial role in ensuring that the team delivers value to the business. They write customer-centric items, typically user stories, prioritize them, and add them to the product backlog. This empowers the Scrum teams to focus on the most critical tasks.

Production Kanban – A signal specifies the type and quantity of product an upstream process must produce.

Program – A program is the umbrella structure established to manage a series of related manufacturing projects. The program does not produce any manufacturing project deliverables. The manufacturing project teams produce them all. The purpose of the program is to provide overall direction and guidance, to make sure the related manufacturing projects are communicating effectively, to provide a central point of contact and focus for the client and the manufacturing project teams, and to determine how individual manufacturing projects should be defined to ensure that all the work gets completed successfully.

Program Manager – A program manager is the person with the authority to manage a program. (Note that this is a role. The program manager may also be responsible for one or more of the manufacturing projects within the program.)

Pull – Material flow triggered by actual customer need rather than a scheduled production forecast. Downstream process signals to upstream processes precisely what is required and in what quantity.

Push – The production of goods regardless of demands or downstream needs, usually in large batches to ensure "efficiency." This strategy, while common, may not always be the most efficient or customer-centric approach to production.

Quality % – refers to the quality of the products currently being produced by a machine. Manufacturers rely on a set of metrics to determine a product's quality. With manufacturing analytics, it's possible to calculate a part's quality by collecting part count and reject reason codes and augmenting this information with input from human operators.

Quality Function Deployment (QFD) – A systematic process used to integrate customer requirements into every aspect of the design and delivery of products and services. Output graphics are often the House of Quality.

R Charts – Plot of the difference between the highest and lowest in a sample Range Control Chart.

Random – Selecting a sample so each item in the population has an equal chance of being selected, but there is a lack of predictability.

Random Cause – A source of variation which is random; a change in the source ("trivial many"), e.g., a correlation does not exist; any individual source of variation results in a small amount of variation in the response; cannot be economically eliminated from a process; an inherent natural source of variations.

Random Effects Model – Experimental treatments are a random sample of a larger population of treatments. Conclusions can be extended to the population. Interferences are not restricted to the experimental levels.

Random Sample – One or more samples randomly selected from the universe (population).

Random Variable – A variable that can assume any value of a set of possible values.

Random Variation – Data Variations resulting from causes that cannot be pinpointed or controlled.

Randomness – A condition in which any event in a set of events has the same mathematical probability of occurrences as all other events within the specified set, i.e., individual events are not predictable even though they may collectively belong to definable distribution.

Range – The difference between the highest and lowest in a set of values or "subgroup."

Ranks – Values assigned to items in a sample to determine their relative occurrence in a population.

Ratio – A numerical scale has an absolute zero point and equal units of measure based on measurements of an output parameter, e.g., amps.

Regression Analysis – includes any techniques for modeling and analyzing several variables. Linear regression was the first regression analysis to be studied rigorously and used extensively in practical applications.

Reject Region – The region of values of which the alternate hypothesis is accepted.

Repeatability (of a Measurement) – The extent to which repeated measurements of an object with a particular instrument produce the same value.

Replication – Observations made under identical test conditions.

Representative Sample – A sample accurately reflects a specific condition or set of conditions within the universe.

Reproducibility (of a Measurement) – The extent to which repeated measurements of an object with a particular individual produce the same value.

Requirements – are descriptions of how a product or service should act, appear, or perform. Requirements generally refer to the features and functions of the deliverables you build on your manufacturing project. Requirements are a part of the manufacturing project scope. The high-level scope is defined in your manufacturing project definition (charter). The requirements form the detailed scope. After your requirements are approved, they can be changed through the scope change management process.

Research – Critical and exhaustive investigation or experimentation, aiming to revise the accepted conclusion in light of newly discovered facts.

Residual Error – See Experimental Error.

Response Time – The Time that an order needs to be satisfied.

Retrospective – A team meeting at the end of every iteration to review lessons learned and discuss how the team can be more efficient in the future. It is based on applying what was learned from the previous sprint to the upcoming sprint.

Rework – Activity required to correct defects produced by a process.

Risk – Potential external events may negatively impact your manufacturing project. Risk refers to the probability of the event occurring and its impact on the project if it occurs. Suppose the combination of the likelihood of the occurrence and the effect on the project is too high. In that case, you should identify the potential event as a risk and put a proactive plan in place to manage the risk.

Robust – The conditions or state in which a response parameter exhibits hermetic to external cause of a nonrandom nature, i.e., impervious to perturbing influence.

Safety Stock – Inventory held to compensate for variations in demand, quality, and downtime.

Sample – One or more observations drawn from a more extensive collection of observations or universe (population).

Scatter Diagrams – Charts that allow the study of correlation, e.g., the relationship between two variables.

Scope – Scope is how you describe the manufacturing project's boundaries. It defines what the manufacturing project will and will not deliver. The high-level scope is set in your manufacturing project definition (charter) and includes all your deliverables and the boundaries of your manufacturing project. The detailed scope is identified through your business requirements. Any changes to your manufacturing project deliverables, boundaries, or requirements would require approval through scope change management.

Scope Change Management – Scope change management aims to manage changes to previously approved scope statements and requirements. The scope is defined and approved in the scope section of the manufacturing project definition (charter) and the more detailed business requirements. Suppose the scope of the business requirements changes during the manufacturing project (and usually, this means that the client wants additional items). In that case, the cost, effort, and duration estimates may no longer be valid and therefore require scope change management.

Scrum – Scrum is a framework within which people can address complex adaptive problems while productively and creatively delivering products of the highest possible value. It is based on the adaptive and iterative methodology.

Scrum Master – Scrum is accountable for removing impediments to the ability of the team to deliver the sprint goal/deliverables. The ScrumMaster is not the team leader but acts as a buffer between the team and any distracting influences. The ScrumMaster ensures that the Scrum process is used as intended. The ScrumMaster is the enforcer of rules. A key part of The Scrum Master's role is to protect the team and keep them focused on the tasks. The role has also been called servant-leader to reinforce these dual perspectives.

Scrumban – Scrumban is a mix of Scrum and Kanban that supposedly contains the best features of both methods.

Sensei – Japanese word for "teacher" and denotes mastery within their field of knowledge. A Sensei should be a wise and easily understood mentor who guides thinking with his subjects rather than dictates the point to promote learning.

Setup Time – The time required to change a process after producing the last part of one product to the first good part of the following product.

Short Term Variation – The amount of variation observed in a characteristic that has not had the opportunity to experience all the sources of variation from the inputs acting on it.

Signal Kanban – A signal that triggers an upstream process to produce when a minimum quantity is reached at the downstream process.

Single Minute Exchange of Die (SMED) – A technique to reduce setup or changeover times to eliminate the need to build in batches.

Spaghetti Chart – A visual chart showing the path taken by a product or person during a process to highlight excessive motion.

Specification Limits – The boundaries of acceptable performance.

Spike – A short, time-boxed piece of research, usually technical, on a single story that is intended to provide just enough information that the team can estimate the size of the story.

Sponsor (executive sponsor and manufacturing project sponsor) – The sponsor is the person who has ultimate authority over the manufacturing project. The executive sponsor provides manufacturing project funding, resolves issues and scope changes, approves major deliverables, and provides high-level direction. They also champion the manufacturing project within the organization. Depending on the manufacturing project and the organizational level of the executive sponsor, they may delegate day-to-day tactical management to a manufacturing project sponsor. If assigned, the manufacturing project sponsor represents the executive sponsor daily and makes most decisions requiring sponsor approval. If the decision is large enough, the manufacturing project sponsor will take it to the executive sponsor.

Sprint Backlog – The team plans each sprint at the beginning, resulting in a backlog of work that the team anticipates completing at the end of the sprint. These are the items that the team will deliver throughout the sprint.

Sprint Planning – This is a pre-sprint planning meeting attended by the core agile team. During the meeting, the Product Owner describes the highest priority features to the team as described in the product backlog. The team then agrees on the number of features they can accomplish in the sprint and plans out the tasks required to deliver those features. The planning group works on the features of User Stories and assigns Acceptance criteria to each story.

Sprint Review – A Sprint review follows each Sprint. During this review, the software developed in the previous Sprint is reviewed, and new backlog items are added if necessary.

Sprint/Iteration – A fixed duration period where user stories are chosen to work on. Sprint comes from the Scrum methodology and is analogous to the term iteration. A sprint is defined as a 2–4-week.

Stable Process – A process free of assignable causes, e.g., in statistical control.

Stakeholders – Specific people or groups who have a stake in the outcome of the manufacturing project are stakeholders. Usually, stakeholders are from within the company, including internal clients, management, employees, administrators, etc. A manufacturing project can also have external stakeholders, including suppliers, investors, community groups, and government organizations.

Standard Deviation – One of the most common measures of variability in a data set or in a population. It is the square root of the variance.

Standardized Work – A defined work method that describes the proper workstation and tools, work required, quality, standard inventory knacks, and sequence of operations.

Statistical Control – A quantitative condition that describes a process free of assignable/special causes of variation, e.g., variation in the central tendency and variance. Such a condition is most often evidenced on a control chart, i.e., a control chart that displays an absence of non-random variation.

Statistical Process Control (SPC) – Use of basic graphical and statistical methods for measuring, analyzing, and controlling the variation of a process to improve the process continuously.

Steering Committee – A steering committee is usually a group of high-level stakeholders responsible for guiding overall strategic direction. They don't replace a sponsor but help spread strategic input and buy-in to a more significant portion of the organization. The steering committee is especially valuable if your manufacturing project impacts multiple organizations because it allows input from those organizations into decisions that affect them.

Story Points – Unit of estimation measuring complexity and size.

Subgroup – A logical grouping of objects or events that displays only random event-to-event variations, e.g., the objects or events are grouped to create homogenous groups free of assignable or special causes. By the minimum within-group variability, any change in the central tendency or variance of the universe will be reflected in the "subgroup-to-subgroup" variability.

Supermarket – A strategically controlled store of parts used by downstream processes.

Supplier – vendor on the entity responsible for providing input to a process through resources or information.

Symptom – That which serves as evidence of something not seen.

System – That which is connected according to a scheme.

Systematic Variables – A pattern that displays predictable tendencies.

Takt time – is the rate of customer demand. It is the available operating time for the requirement. More simply, it is the availability of service time divided by the customer demand rate. Takt time enables your organization to balance the pace of its service outputs to match the rate of customer demand.

Task – A user story can be divided into one or more tasks. The developer working on them estimates the tasks remaining daily in hours (or story points).

Task board/Storyboard – A wall chart with cards and sticky notes representing all the work for each sprint. The notes are moved across the board to show progress.

Team – The Team is responsible for delivering the product. A Team typically comprises 5–9 people with cross-functional skills who do the actual work (Analyze, design, develop, test, technical communication, document, etc.). It is recommended that the Team be self-organizing and self-led, but they often work with some form of manufacturing project or team management.

Test of Significance – A procedure to determine whether a quantity subjected to random variation differs from the postulated value by an amount more significant than that due to random variation alone.

The Manufacturing Kaizen – A rapid improvement event; part of a continuous improvement program. A focused, dedicated, and well-defined event that is used to get quick hit value by implementing "do-now" solutions leading to waste elimination.

The Theory of Constraints – describes the methods used to maximize operating income when an organization is faced with bottleneck operations. This theory also deals with handling the unknown.

Theory – A plausible or scientifically acceptable general principle offered to explain phenomena.

Time Boxing – Time boxing is a planning technique common in manufacturing projects. In this technique, the schedule is divided into several periods (time boxes, usually 2–6 weeks long), with each part having its deliverables, deadline, and budget.

Time Study – A time study analyzes every step of the manufacturing process to determine, on average, how long each step takes.

Total Cycle Time (TCT) – The time taken from work order release into value stream until completion/movement of product into shipping/finished goods.

Total Productive Maintenance – Maximizing production system efficiency by analyzing and eliminating downtime through up-front maintenance.

Toyota Production System – Toyota Motor Company's production system focuses on eliminating waste throughout the value stream.

Trend – A gradual, systematic change over time or some other variable.

Two-Sided Alternative – The value of a parameter that designates an upper and lower bound.

Type I Error – See Alpha Risk.

Type II Error – See Beta Risk.

Unnatural Pattern – Any pattern in which many measurements do not group themselves around a center line; when the pattern is unnatural, outside disturbances are present and affect the process.

Upper Control Limit – A horizontal line on a control chart (usually dotted) representing the upper limits of process capability.

User Personas – Personas describe the typical users of a given software and should include skills, background, and goals.

User Story – A user story is a very high-level definition of a requirement containing just enough information so that the developers can reasonably estimate the effort to implement it. It is one or more sentences in the everyday capture of what the users want to achieve. A user story is also a placeholder for conversation between the users and the team.

Utilization – often used in conjunction with capacity, "capacity utilization" is a percentage or KPI showing the total capacity being used at a particular time.

Value – refers to a product or service capability provided to a customer at the right time and at an appropriate price.

Value-Added Activity – Any activity that changes the product's fitness, form, or function toward something a customer is willing to pay for.

Value-Added Time – The time spent in value-added activity to produce a unit. It is time for those work elements to transform the product into something that makes the customer willing to pay.

Value Stream – All value-added and non-value-added activities required to bring a product, group, or service from the point of order to the hands of a customer and a design from concept to launch to production to delivery.

Value Stream Map – A visual representation of a process showing the information and material flow through all steps from the supplier to the customer.

Variable – A characteristic that may take on different values.

Variables Data – Numerical measurements made at the interval or ratio level; quantitative data, e.g., ohms, voltage, diameter, and subdivisions of the measuring scale, are conceptually meaningful, e.g., 6,478 V.

Variation – Any quantifiable difference between individual measurements. These differences can be classified as being due to common causes (random) or unique causes (assignable).

Variation Research – Procedures techniques and methods used to isolate one type of variation from another (for example, separating product variation from test variation).

Velocity – is a relative number that describes how much work the team can accomplish over a period.

Visualization – The workplace design should allow problems and issues to be identified without timely and in-depth investigation. Indeed, visual workplaces should be capable of assessment in less than 3 seconds.

VOB – The voice of the business is derived from financial information and data. The voice of the company represents the business's needs and key stakeholders. It usually includes profitability, revenue, growth, market share, etc.

VOC – Voice of the customer which represents the expressed and non-expressed needs, wants, and desires of the recipient of a process output, a product, or a service. It is usually described as specifications, requirements, or expectations.

VOE – Voice of the employee. Represents the expressed and non-expressed needs, wants, and desires of the employee to be successful.

VOP – Voice of the process, representing a process's performance and capability to meet business and customer needs.

Waste (Muda) – includes anything that does not add value to a final product or service, an activity that consumes valuable resources without creating customer value.

WIP (Work in process) – These are items – material or information – between machines, processes, or activities waiting to be processed; any inventory between raw materials and finished goods.

Withdrawal Kanban – A signal specifying the type and quantity of product the downstream process may withdraw.

Work Cells – An arrangement of people, machines, materials, and methods such as processing steps are adjacent and in sequential order; thus, parts can be processed one at a time.

Work Plan (schedule) – The manufacturing project work plan tells you how you will complete the manufacturing project. It describes the activities required, the sequence of the work, who is assigned to the job, an estimate of how much effort is needed, when the work is due, and other information of interest.

X – Input

X&R Charts – A control chart representing process capability over time displays variability in the process average and range over time.

Y – Output

Index

Note: *Italic* page numbers refer to figures.

FSC
www.fsc.org
MIX
Papier | Fördert gute Waldnutzung
FSC® C083411

Kolibri 360 GmbH
Ferdinand-Jühlke-Straße 7
99095 Erfurt, Deutschland
produktsicherheit@kolibri360.de